Our Changing Geopolitical Premises

Our Changing Geopolitical Premises

THOMAS P. RONA

Transaction Books
New Brunswick (U.S.A.) and London (U.K.)

Library of Congress Catalog Number: 81-16192
ISBN: 0-87855-897-7
Printed in the United States of America

Library of Congress Cataloging in Publication Data

Rona, Thomas P.
  Our changing geopolitical premises.

  Bibliography: p.
  1. United States—National security.
2. World politics—1975-1985.   3. Twentieth century
—Forecasts. I. Title.
UA23.R67   355'.033073   81-16192
ISBN 0-87855-897-7 (pbk.)   AACR2

To the Memory of Ben T. Plymale

He was the man who did the most to inspire this book; many of his views found their way into its message. He did not live to see its publication. It is perhaps not inappropriate to offer this work as a modest tribute to his memory.

Others will praise his expertise in military matters of vital import to our nation, his broad insight ranging all the way from the intricacies of advanced technology to the lofty levels of national policy formulation. In the heart of his close associates, he will be remembered as a man, a friend, a leader and as an example for all of us to follow.

We will long remember his phenomenal capacity for painstaking work; always asking more from himself than from any of his co-workers, even as in the course of the last years his health was failing. He was willing to accept and to nurture the ideas of others, generous in assigning credit and in forgiving honest mistakes. Not for a moment could anyone entertain any doubt as to his standing with him; his language was forceful, pithy and direct. He had no patience with phonies, fainthearted or fools; compromise on matters of principle was definitely not for him.

Long will his memory live among all of us.

ACKNOWLEDGMENTS

The views expressed in this book are those of the author. They should not be construed as reflecting the positions of any corporation or government agency.

Grateful acknowledgment is made to the following for allowing reprint permission:

Figure E-1, page 281, reprinted by permission of *The Wall Street Journal*, © Dow Jones & Company, Inc., 1980. All rights reserved.

Photograph, page 96, by Ann Yow, February 24, 1980 edition of *The Seattle Times*.

Cartoon, page 33, by Herblock, from "Herblock On All Fronts" (New American Library, 1980).

Cartoon, page 151, *Die Ziet*, Nr. 17, April 1980. Reprinted by permission.

Appendix Note C, pages 270–76, "Anatomy of a Crisis—The DC-10 Accident," contains material from text obtained from The Information Bank, New York Times Company. Reprinted with permission.

CONTENTS

Foreword

That work of this nature could be pursued under the auspices, and with the active support, of senior managers of the U.S. aerospace industry is a tribute to their increasingly sophisticated perception of the true U.S. national requirements in the broad, long-term geopolitical context.

In attacking a subject as multifaceted as the problem of protecting the long-term U.S. national security interest beyond the single military dimension, one quite naturally starts with some personal biases. I freely confess to mine, which are centered on the idea that the politically and technologically advanced societies, as a group, should find a *modus vivendi* with their adversaries based on something other than the threat of mutual destruction backed by the tenuous hope that some measure of differential recovery capability may designate the "winner" in a major nuclear exchange. I also propose to explore the thought that many, and perhaps most, of the threats to the future way of life of the industrially advanced nations may not be due to the efficiently nefarious communist designs, but to political and economic problems that are likely to confront them all, including perhaps, in the more distant perspective, even the Soviet Union.

If the causes of the almost endemic state of conflict, which seems to infect the entire world and threatens at any time large-scale destruction of the human societies as we know them, are correctly identified, then perhaps remedies can be found and eventually applied. This is a task of no inconsiderable magnitude, to say the least; never in human history has it been attempted, let alone brought to successful conclusion. But even if all the insights are complete and all the perceptions correct; even if all judgments are wise and decisions competent—even then, the implementation process is likely to take several decades; perhaps more than one generation. This time delay is the crucial element: the essential determinants of the problem evolve much faster than the lineaments of attractive solutions. In the meantime, the visions of impending Apocalypse blur our vision and even further detract from whatever rationality might still be present.

The key to our future security then is to deter (and thereby defer) military conflicts at all levels by using a broad spectrum of economic and political, as well as military, deterrent elements until the currently antagonistic major power centers acquire more vested interest in the protection of the peaceful status quo than in the advocacy or pursuit of the dangerous path of violent confrontations.

In order to maintain the required military strength on a permanent, long-term basis, the Western nations must, in concert, establish the ideological, social, and economic matrix, broadly attractive and acceptable to their constituencies. In relation to its current antagonists, it becomes quite apparent (to the surprise of many) that the West may, with proper forethought, marshal its immense resources to provoke a "correlation of forces" of its own, working in support of its own broadly perceived interests, as contrasted to reacting passively and sometimes irrationally to hostile initiatives.

Lest I lose all credibility with my reader (who, in my fond hopes, is sincerely interested, well-informed, perhaps also indulgent, mostly patient—but certainly not foolish or naive), I might as well avoid right here any impression that the problem approached here is "solved" in any measure, or even that *the* major component of the "solution" is fully and competently described. I attempt to offer a hypothesis, perhaps leading to future explorations, which, if conducted with dedication and objectivity, may be eventually recognized as modest steps in the right direction.

I record with pleasure and gratitude the unfaltering support of The Boeing Company. For many years it has provided the proper environment and the many stimulating associates and challenges which are prerequisites for this type of undertaking. The purely material aspects of this support should not be underestimated; just as the pen is often favorably compared with the sword, no less than a modern word processor is required when attempting to confront the causes of mortal conflicts of the future.

The pleasurable and often essential interactions with professional associates, many of them dear friends, must be clearly acknowledged, even though the specific individual contributions cannot be identified. Credit, for whatever might be found creditable, should clearly go to them; errors of judgment or interpretation are recognized as being my very own. Particular recognition is due to Dr. W. A. Hane who spent innumerable hours pondering my arguments, testing with the *aqua regia* of his logic their strengths while charitably condoning their weaknesses; always tempering with his quiet wisdom my enthusiasm which, as time flies by, cannot even be ascribed any longer to youthfulness. May the record show that this work would not have been conceived without the inspiration of his presence on a certain sun-drenched Mexican beach, under the watchful eye of patrolling pelicans and cavorting cormorants. In a different, but just as important, dimension I wish to acknowledge my deep gratitude to Mrs. Dorothy P. Linder for her patience, devotion, understanding, and faith. Amidst interminable hours of secretarial drudgery, she has participated as a trusted and competent research associate, sharing thus the protracted horrors of intellectual doldrums as well as the short moments of creative elation.

June 1980

POSTSCRIPT

Being prematurely right is said to be dangerous, but seeing one's prophecies gaining widespread acceptance *before* they are published is downright vexatious. This work was undertaken in March 1979 and essentially completed by May 1980. Since that time the pressure of other business and mostly the fascinating evolution of national and world events have offered handy excuses to postpone the final crafting of the conclusions. Whenever appropriate, I avenge myself for having been overtaken by history by adding a few comments identified as "Note added in proof." Otherwise the text remains unchanged.

April 1981

Part I
STRUCTURING THE PROBLEM

CHAPTER I
Introduction

The advanced industrial nations enjoy an unprecedented degree of prosperity if we choose to measure it by the rates of consumption. One may be justified in expecting just about everywhere stable and well-entrenched governments, supported and acclaimed by sizable majorities. According to elementary logic and common sense, individual citizens should exude enthusiastic devotion to their political systems and express wholesome confidence in their own future. In their attitudes, as in their actions, they should demonstrate willingness to preserve and readiness to protect the institutions which have so wisely fostered (if not indeed created) their prosperity.

By inveterate perversity, or perhaps visceral prescience, Western societies taken as a whole and influential segments of their populations apparently fail to respond properly according to such logic. Generally among the major industrialized societies, but more particularly in the U.S., the abundance of material possessions, the spread of conspicuous consumption habits, and the rapidly growing entitlements to social benefits have all failed, individually and collectively, to bring forth broadly based optimism and contentment, let alone an increased sense of individual responsibility toward the common good in the *national perspective*. As to longer term concerns of international or global scope, they are hardly ever mentioned except in the arcane purlieus of the U.N. committees. That they do not, in general, excite vivid debate within the U.S. Government and the public is a delicate understatement. Even in the narrow local perspective, popular expectations seem to outpace by a considerable margin what society can possibly deliver. Subjective disappointments often lead to an almost endemic state of suspicion, bitterness, alienation and hostility, with violence always subliminally implied if not explicitly threatened. Discontent is far from remaining the privilege of the poor (the "underprivileged"); it seems to affect in different forms and degrees just about every segment of society. The upper classes (as measured, in an admittedly arbitrary manner, by annual income levels) vocally resent the trends which decrease their rewards and prerogatives; they believe that these are no longer consistent with their efforts, talents, and willingness to take risks. Furthermore, they feel that their fundamental birthright, or at least motivation, to pass the sources and the rewards of their privileged position on to their offspring is being threatened to the point that it may well be legislated out of existence within their own lifetime.[1]

The lower classes, still as measured by income, and those groups which for whatever reason rightly or wrongly feel the effects of discrimination, most easily and willingly take the mental transition steps from fairness, equal rights, equal opportunity, and equal protection to the equality of rewards, quite irrespectively of individual merits or efforts. Shrill voices can be heard in support of "reparations" for injustices in the long distant past; playing on not-so-subliminal collective guilt feelings of the public, and with the help of often strained legal arguments, the groups allegedly victims of such injustices actually manage to exact reparations of this type from the larger community.

That all this, plus many other instances I could enumerate, amounts to a rather widespread *malaise*, growing disenchantment toward, and pervading cynicism about, our political system can no longer be questioned. That the electorate responds by individual disaffection and by the increasing prevalence of legally organized pressure groups is

---

[1] Much has been written, probably with more feeling than factual justification, on the "proletarization of the middle class" in Great Britain, for instance. The recently elected, strongly conservative, Thatcher government appears to be sincerely committed to reversing this trend. It is obviously premature to assess at this point the probability of success.

a fact supported by observation. Government reaction by confusion, tergiversation, procrastination, by groping for short-term "muddling-through" type solutions in a quasi-permanent crisis atmosphere has by now also become the familiar backdrop of the political scene. And notice that this signal failure of the governments to provide true, insightful leadership, as contradistinguished from doctrinary pursuit of atavistic party principles, is not peculiar to the U.S. system; it is, with only minor variations, present in most of Western Europe and Canada as well. Only Japan appears to have been able, at least as seen from the relatively distant U.S. vantage point, to select governments which can reach decisions with regard to longer term goals and implement the corresponding action policies with a reasonable degree of public consensus.

The failure of so many governments, all claiming power from the consent of the governed, in their essential task of perpetuating public commitment to the prevailing political institutions seems to be a widespread, monotonically growing phenomenon, clearly perceptible since the first decades of this century. It appears to be, within quite broad limits, independent of national characteristics and it is only temporarily checked or subdued in times of national emergencies. Clearly, it cannot be remedied by the modern version of the *"panem et circenses"* principle of imperial Rome, the monstrous growth of transfer payments, and the vicarious titillations offered by television. We must first, if we are to pursue a rational path leading to remedies, search for the major causes associated with the fundamental or structural features of modern society itself.

In this essay, I propose to identify at least one, possibly major, cause of what could be termed the incipient decay of the social fabric of the Western-style industrialized nations. Starting with the plausibility of such a major cause, I then proceed to examine some of our commonly accepted socio-political ("domestic") and geopolitical ("international") premises in the hope of contributing to the alleviation, or even elimination, of that cause. Not surprisingly, perhaps, I find that some of our important commonly accepted premises do indeed require serious scrutiny; that the premises upon which domestic and international policies are formulated must be revised to make them mutually congruent or at least complementary in the near-term future and well to the limits of our time horizon; and that proper articulation of these revised premises could be of significant help in establishing a viable U.S. foreign policy while still satisfying the domestic-oriented mandates of the Constitution. Almost as an afterthought, I also venture to suggest that we may, as a by-product of this endeavor, find an ethically acceptable, or even emotionally attractive, rationale for the continuing development of the multi-faceted components (political, economic, and military) of our strategic posture. Light might also be thrown on the apparent contradictions between emphasis on negotiated strategic arms limitations and continuing massive investments in nuclear as well as in "conventional" weaponry.

CHAPTER II
The Symptoms

It may be opportune at this point to recapitulate in some sort of a "bill of particulars" the more readily observable symptoms of the incipient societal decay just discussed.

The loss of cohesion, common purpose, and public confidence in our institutions as well as in individuals holding public office at all levels of government has already been mentioned. Opinion polls, to the extent that they reliably reflect public attitudes, substantially corroborate these trends. The participation in the electoral process, as evidenced by the fraction of eligible voters actually casting their ballots, has steadily decreased since the mid-1950's; in some instances the absolute number of votes cast per precinct has actually decreased in spite of the growth in population and of the commendable efforts to extend voting franchise to segments which were heretofore denied this privilege.[1] The participation of minorities in national (as distinguished from local) elections has been one of the major disappointments of those who traditionally champion liberal causes. Added to the nonparticipation is the almost universal absence of true consensus on any issue of substance, let alone enthusiastic endorsement of any candidate who would display the fortitude of taking an unambiguous stand on such issues. Presidential elections hinge on minute shifts in fractional percentages even with candidates described as

---

[1] (Note added in proof) The 1980 Presidential election had the lowest voter turnout since 1952. Only 52.7% of the eligible voters participated, in spite of the many fundamental issues and sensational headlines that should have elicited some measure of involvement by the electorate.

7

"charismatic vote getters"; a few percentage-point majorities are hailed by the press as nothing short of landslides. It takes a luridly emotional issue (e.g., abortion, homosexuals) or a tangible pocketbook issue (e.g., tax roll-back) to bring forth significant majorities. The participation and the consent of the governed are myths that live on mostly in the textbooks foisted upon the unsuspecting, and at any rate defenseless, highschool student.

*

The basic mechanism for self-perpetuation in positions of power, used by both the legislative and executive branches of the Government as well as by individual politicians masquerading as statesmen, is to *cater to the highly visible near-term self interests* of the electoral majority—while ignoring, or even deliberately misrepresenting, the concomitant longer term penalties. The converse process, namely to advocate immediate and tangible sacrifices in order to protect long-term societal interest, is so unusual that it can be safely characterized as a political aberration. The rationale supporting this trend is hardly subject to argument: Since the majority of the voters have low-to-moderate incomes and since the "time horizon of concern" of these voters tends to be short[1], the well-advised politician will ensure his (or her) own longevity by staunchly supporting causes with highly visible near-term payoff. The cost, according to popular myth, is to be paid by those who can afford it--business and wealthy individuals; the day of true reckoning, if given any thought at all, is conveniently relegated to the murky recesses of the future. Illustrative examples abound in the course of the past two decades. Social security, which originally was conceived as offering minimal insurance against failure of individual old-age provisions, has been transformed into a massive (and bankrupt) scheme for income transfer favoring older generations. Health care for the elderly is already part of social security, and national health care for all is strongly advocated in certain quarters as the next milestone of social progress. Welfare payments, with only minimal safeguard against the most blatant abuses, have reached levels which cause most low-paying jobs to look unattractive in comparison. Instead of being aimed strictly at the relief of true and unusual hardship, welfare has become a mechanism for alleviating practically all the penalties society normally inflicts on marginal producers. All this, of course, is in the name of compassion, human dignity and sometimes even equal opportunity. Many other instances could be easily found, but I hope that the basic point is made: Our political system is increasingly biased toward providing immediate, consumption-oriented rewards to a large fraction of the population, without requiring commensurate productive efforts by the recipients. It should go without saying (but it is important to remember) that, compounding the direct economic and ethical impact, this trend also encourages uncontrollable growth of "social overhead" required by planning, administering, and (ineffectually) auditing the process. That this in turn creates self-reinforcing institutional momentum is a safe surmise in any modern democracy.[2]

*

Public attitudes and principles guiding our legislation, while *ostensibly* derived from the Constitution and the Bill of Rights, are in fact distorted, often unwittingly, but sometimes with self-serving deliberateness, so that their net societal impact becomes demonstrably nefarious. The responsibility for this trend must be shared by all three branches of the Federal Government, by their counterparts in the state governments, by many professional and special-interest groups and, perhaps quite significantly, by the academic profession.[3]

---

2 (Note added in proof) The early initiatives and efforts of the newly installed Reagan Administration show a most commendable attempt in the direction of reversing these trends.

3 Activism was the term coined to describe the attitudes of Supreme Court Justices who have chosen to inject, publicly and explicitly, their personally held social convictions into their function of interpreting the Constitution. The judicial "activism" of the U.S. Supreme Court under Chief Justice Warren (1952-69) was to a great extent abetted, if not engendered, by the strong liberal trends in the leading law schools of the country. As long as the Justices are human, their opinions will always be colored by their personal convictions.

Illustrations are plentiful in our post-World War II history. Higher education, being a publicly financed (or at least publicly favored) societal function, has come to be perceived as a right, rather than a privilege. Universities have been (not always or necessarily against their wishes) enticed or pressured into processing a vastly increased number of students, with the results visible to all: Overproduction of graduates with few marketable skills, dilution in the value of university degrees, grade inflation, and (not the least) loss of purpose among many young people.

In the name of freedom of speech, the public information and entertainment media offer a vocal forum to manifestations of all sorts of social deviancy. Paraphrasing Gresham's Law, one might conclude that the socially worthless, or even positively harmful, messages and publications tend to drive out those which are valuable and beneficial.   Thus, convicted criminals, not exactly illustrious on the basis of their past literary achievements, can easily earn a fortune by just lending their name to ghost-writers selling hastily concocted "memoirs." Billions are being spent each year on commercial advertising, just to achieve minute shifts in brand preferences among almost identical products[4], some of which have been furthermore proved harmful, according to the best available evidence. Sordid purveyors of pornography, with meretricious support by members in good standing of the legal profession, seriously argue in court the artistic merits of the full-color representation of the naked human body, without discrimination on basis of sex but with meticulous, if tasteless, attention to the genitals.   It becomes obvious that the sensation value of deviancy and the greed for business profits by far outweigh the ethical self-restraints which should be preconditional and concomitant to free speech in a well-regulated society.

Misconstruction of Constitutional tenets mostly imposed, or at least reinforced, by judicial activism is best illustrated by the recent history of the "equal protection under the law" principle.   What originally started as a genuine attempt to eradicate *legal* discrimination against any group or indi-

---

[4] Except "image" and packaging, of course.

vidual, has by now grown into an intricate tangle of legislation. It is reinforced by a multitude of administrative and judicial decisions, aimed at penetrating the most minute, not to say private, aspects of our individual behavior. Nondiscrimination in education has grown into city-wide busing and pupil assignment; in housing, it has disrupted neighborhoods through judicial nullification of local zoning laws. Equal opportunity in employment has come to be translated into quotas favoring women and racial minorities even in areas where bona fide preferences may be amply justified. [5] On whatever endeavors involve public monies or tax preference to the slightest extent (and it is hardly possible to find any nowadays where they do *not*), the heavy hand of nondiscriminatory regulation has descended. The result may be seen as a Government-imposed effort to amalgamate, or even homogenize, the culturally diverse groups in our society, not in any way mandated or implied by the Constitution—and certainly not with the consent of the governed.

Still under the heading of misconstruction of Constitutional tenets, one may mention the excessive emphasis on the accused's civil rights (all in the name of "equal protection") which has slowed down the process of criminal justice and has, for practical purposes, increased the safeguards of the criminal's rights and the cost, if not the chances, of rehabilitation at the expense of the protection of the victims.

One should not place the full burden of responsibility for the trend we discuss here on the Government. The population, when not guided by those responsible for shaping and articulating public attitudes, must definitely bear its share. As an illustration, let us consider what has recently happened to the "pursuit of happiness," so clearly stated by the Declaration of Independence as one of the (permissive) purposes of Government. It was always more or less agreed that individual efforts in the pursuit of happiness must be constrained at the limits where they start impinging

---

[5] It has taken not less than a U.S. Supreme Court ruling to decide (June 1979) that a completely deaf person is not *entitled* to admission in a graduate nursing school.

adversely on similar efforts by others. In recent years, however. the "do your own thing" ethic, first advocated by the younger groups in the late 1960's, has made detectable inroads on other segments of society. Taken literally, such ethic endorses the onesidedness of the social contract: The individual is encouraged to enjoy the benefits of society now and in the future, but with no obligation whatever to contribute to the stability and the protection of that society. A strong and affluent group may support a small minority holding such views; it may even benefit from the ferment introduced by the attendant intellectual diversity. But it should be evident that no society can afford high prevalence of such luxury in the long run without very real threat of internal decay—this grows even more ominous when society becomes subjected to external pressures.

The misconstruction or the distortion of our Constitutional principles, by the Government and by the public, is bound to erode and to fracture the "ideological cement" that ties individuals together in a purposeful society.

* * *

The U.S. foreign policy has failed in the past twenty years to follow coherent long-range themes consonant with our asserted political ideals; it cannot even be praised (or excused) on the basis that at least it has promoted success- fully our narrowly construed economic interests. In broad summary terms, our foreign policy has resulted in: (1) the erosion, through inaction and negotiations, of our once unquestioned global military supremacy; (2) weakening and even disruption in some cases of the U.S. alliance structure[6]; (3) the loss of national will, or the military preparedness (or both), to oppose hostile peripheral events, even when these clearly threaten our interests that once would have

6 The basic dichotomy in the relationships with our major Allies is not being explicitly recognized in the conduct of our foreign policy. We must encourage and support their prosperity and security by tangible actions, including (but not limited to) shouldering a disproportionate defense burden justified only on the basis of long-past differences in disposable

been called vital, or when they are convincingly abetted by endeavors hostile to the U.S.; (4) the loss of eminently advantageous negotiating positions in matters of international monetary and trade agreements; and (5) the failure to define a consistent and sound rationale for our behavior toward the less developed nations capable of ensuring long-term stable political and economic relationships while still providing access on reasonable terms to raw material supplies so strongly associated with our security and well-being.

Significantly, when adverse trends become publicly apparent or when discrete events temporarily shatter public complacency, the U.S. leadership—Administration and Congress conjointly—seems to affect ignorance or, when that is not possible, explains them away.[7] Convenient and ad-hoc rationalizations for confusion, inaction, or appearance of action are usually found and, when necessary, fabricated. The U.S. public is generally oblivious to foreign events until some domestically perceptible impact jumps into the headlines. But, to the extent that it follows and assesses the results of our recently pursued foreign policies, it receives little hope and encouragement for the long-term prosperity of the U.S. And, worst of all, there is nothing on the horizon to suggest significant changes in our approach to the rest of the world.

6   (Footnote continued)
    national wealth. At the same time, the very growth in relative prosperity of our partners contributes to their increasing independence and assertiveness with respect to the conduct of U.S. foreign policy. They also compete rather vigorously and quite successfully for markets and sources of raw materials with their U.S. counterparts. At least in some cases, such as the Allied sales of advanced weapons in Latin America or the opposition to U.S. efforts to limit nuclear weapon proliferation potential through restrictions on nuclear powerplant exports and on fuel reprocessing, the military/economic policies of our Allies have directly countered those pursued by the U.S.

    (Note added in proof) Owing to its rapid onset and highly publicized manifestations, the Islamic revolution of Iran (October 1978 through February 1979) has impinged upon the U.S. domestic political climate to an extent far out of proportion with its intrinsic significance.

7   Written in May 1979, before the "Cuban Combat Brigade" episode in September.

CHAPTER III
The Cultural Misadaptation Hypothesis

Quite naturally, such brutally critical listing of symptoms tends to exaggerate the negative aspects, since our purpose is to identify indications of societal dysfunctions. A balanced analysis would dwell probably at length on the positive side of the ledger, the blessings and benefits so bountifully provided to so many people. In fact, most observers would agree that until the mid-1960's the results of the U.S. political system could be easily described as a remarkable success story. For almost 200 years it has unquestionably served the nation well even in times of rapid and unforeseeable political, economic, and technological changes. It has admittedly benefited from highly favorable initial conditions, but it has also attracted and accommodated under conditions of relative prosperity a massive, heterogeneous influx of newcomers and, following trivially short delays, has extended to them the full rights and benefits of citizenship.

When a nation has for so long shown signs of vigorous growth, it is reasonable to surmise that, in a broad sense, it is well *adapted* to its environment. This concept of adaptation, borrowed from biology, means that the physical and human resources are available and that the means to apply these resources to the interactions with the environment are appropriate to make the net result of these interactions beneficial to the national community.

In order to be useful to our inquiry, this definition of adaptiveness must be carefully qualified. There are many subtleties hidden in its use of deceptively simple terms.

Since our conclusions hinge to a great extent on the careful delineation of the definitions, a few reminders of a somewhat tutorial flavor are offered in Appendix[1] with due apologies to previously initiated readers.

We come now to one of the important assumptions underlying the conclusions presented here:

> The social and political institutions of a nation (the "sociopolitical system") constitute the organizational framework whereby the available and potential resources can be purposefully developed and applied.

Implied in this assumption is the notion that the application of resources will be to the objective (as contrasted to the perceived) benefits of the nation, at least in a well-adapted nation. The term "organizational" is used in the sense of defining functions and rules of interaction; the word "framework" is inserted to remind us once again that detailed individual behavior is not necessarily prescribed.

An assumption of such a sweeping nature and (as I hope presently to convince the reader) of such interesting consequences should not be made lightly without offering plausible arguments in its support. To put Darwin into a nutshell, adaptiveness made possible by the conjoined and opposite effects of mutations and "natural" (environmental) selection is the basis for the long-term survival of the species (breeding group).[2] Subsequent workers in the field have essentially confirmed, and later extended, the concept of adaptiveness to *cultural* aspects as well.

It is commonplace by now to observe that cultural adaptation (although not possible without the genetic equipment) is far more rapid and, thus, far more flexible for relatively short time spans than genetic adaptation alone. The fantastic success of the human species with respect to all others is

---

[1] Appendix - Note A.

[2] Of course, this statement was not formulated in these terms until almost a century after Darwin's seminal publication in 1858.[2]

certainly no coincidence.[3] In terms of adaptation rate and flexibility (at least for the short term) the cultural mechanisms rendered possible by the modern human brain serve as multipliers (amplifiers and accelerators) of man's genetic potential.

With our previous definition of adaptation, the cultural factors are seen to allow more efficient use of resources than the purely genetic ones. Cultural adaptiveness has served human breeding groups remarkably well: The evolving behavioral patterns in familial and tribal groups have ensured favorable differential survival rates; in the later stages, at the communal and regional levels, the benefits of social organization and cooperation soon became equally obvious. Large, successful human groups, conveniently called nations, have enshrined the organizational ground rules into sociopolitical institutions. While the rate of genetic adaptation must be measured over hundreds of thousands of years, cultural evolution has resulted in perceptible, indeed radical, changes in the interaction of human groups with their environment in the last few thousand years.[4]

If we believe that cultural adaptation is the essential mechanism whereby humans have outdistanced their biological competitors; if furthermore we believe that the sociopolitical institutions are extensions and aggregates of the cultural behavior patterns of individuals and smaller groups; then the suggestion that these *institutions are in fact part*

---

[3] The fact that preponderant among animals which offer serious competition to humans are *social insects* is significant. Just so is the fact the leading sociobiologists of our age have metamorphosed into their present advocation from their previous training, and indeed eminence, as *entomologists.*[3] The genetically conditioned effectiveness of a social group (bumblebees) in exploiting the available resources under variable climatic constraints has received considerable attention in recent years.[4]

[4] The recently observed evolution and proliferation rate is no accident. We are not yet at the point where we can purposefully manipulate our genetic equipment, but we can, and most certainly do, powerfully augment our genetically inherited cerebral aptitudes by means of auxiliary artifacts. More about this in Chapter IV.

of society's adaptive mechanisms and thereby constitute the organizational framework which permits the efficient use of resources becomes plausible. It follows that the degree of adaptation of a nation (provided that the intrinsic resources are available) is primarily conditioned by its sociopolitical system. A potentially wealthy nation can unwisely squander its resources and promptly fade from the forescene of history; nations with moderate physical resources can show (and have shown in the past) considerable adaptive stamina and continuing success.

The converse appears also plausible: If a nation, heretofore prosperous, with no catastrophic discontinuity in the availability of its physical or human resources, exhibits within a relatively short time period high incidence of social stresses or fails to respond appropriately to external stimuli, one might be justified in concluding that its socio-political system is inadequate. Since it *was* successful before, the inadequacy must result from a lack of adaptation. More specifically, a given society may simply not detect in time, or not detect at all, changes in the internal or external environment. Once perceived, such changes may be still misconstrued or distorted; the analysis and decision formulation which eventually lead to societal reaction may not be appropriate. Even the implementation of a logical and timely reaction may be precluded or delayed by conflicts with existing beliefs and institutions, by lack of leadership or resources, or by interference by competitive (external) groups.

If the stresses in modern industrialized societies result mostly from their sociopolitical institutions being gradually misadapted, then we should examine the hypothesis that a number of important environmental changes either *have* occurred in the recent past or are *now in the process of unfolding*. Once again, we select for this examination the U.S. as the principal locale, but generalizations to Western Europe, to the white-populated parts of the British Commonwealth, to Japan, and in the more distant future, to Brazil and the Soviet Union, are probably justified.

The first major environmental shift (Chapter IV) is a consequence of the abrupt changes which have occurred in the information flow which reaches individual members of modern societies. This is a typically technological change which could be ignored on the grounds that there are many changes just as abrupt in, say, transportation, energy conversion, material sciences, etc. But the role of the information flow is (and always was) such an essential element in social cohesion and rate of evolution that its being selected for discussion is perhaps not unwarranted.

The second topic chosen for discussion (Chapter V) is an example of a shift in the industrialized societies' *internal* environment; i.e., in the attitudes and self-images which members of these societies develop toward their local and national communities. It can not easily be traced back to simple causes, nor can major milestone events be identified; the shift is nonetheless very much present, widespread, and thought to be consequential of the complexities and rapid rates of change in the relationships between individuals and groups, associated to an increasing extent with the evolution of "post-industrial" societies.

Chapter VI deals with the change (or, rather an increased understanding of such change) in modern societies' *external* environment; i.e., the interactions between human groups, the biosphere, the prevailing physical conditions, and the available material resources. The scale of human economic activity, taken as a whole, is probably still negligible in comparison with that of the geophysical phenomena affecting the Earth on the average, but in a number of local instances the effects of concentration and magnification lead to perceptible, and in some cases irreversibly deleterious, consequences.

The change in the global economic system is the subject of Chapter VII. The economic basis of the rapid rise of industrialized nations rested to a large extent on the beneficial interactions with (some would say, exploitation of) the less developed regions. For many reasons, including ideological competition and erosion of effective military power, the mechanisms of these interactions have by now disappeared or at least have been gradually modified to the point of becoming unreliable.

Chapter VIII surveys the prospects for the use of military power by Western nations for the purpose of expanding or preserving their economic interests. The West appears to operate under constraints which, for the moment, do not apply to its major ideological competitors. This situation is so new in Western historical experience that the principles which may lead to successful adaptation are still being explored.

A detailed discussion of these environmental changes, of their causes and of their impact, is beyond my intended scope. Only a few categorical changes have been selected with some illustrative specifics. Even so, these categories are interacting and overlapping; no originality or uniqueness can be claimed for such grouping, neither is the list intended to be exhaustive. It strongly suggests, though, further and more intensive and detailed search for other major shifts in our environment.

The criteria used to select the changes for discussion are that they be pertinent to modern (mature) industrial societies[5] and that both the onset and the impact of the changes be discernible within a time span of 50 years centered on the present. Little effort was spent to identify all the causal relationships among the changes discussed; these are, as always in sociological matters, complex and largely unexplored. When explored, they turn out to be disappointingly inextricable. In the same spirit, no attempt is being made to separate trends from events[6]; the distinction does not seem to affect the conclusions.

[5] The term "post-industrial" is often used to describe societies where the production potential of industrial goods and services exceeds the unstimulated consumption level.

[6] A discrete event occurs when the rate of onset is rapid compared to the means of perception. A fast trend can be perceived, but evolves faster than the adaptive response of society. Finally, a slow trend is one that evolves at the same time scale as the corresponding adaptive response *if the trend has been correctly perceived*.

Part 2
CHANGES IN ENVIRONMENT

CHAPTER IV
Information Environment: Intensity and Pervasiveness

THE BASIC TOOL OF SOCIAL BONDING HAS UNDERGONE DRASTIC TECHNICAL CHANGES: THE INFORMATION FLOW WITHIN AND AMONG SOCIETIES HAS AND WILL BECOME MORE INTENSIVE AND INCREASINGLY PERVASIVE.

A short historical overview should help to place this assertion in perspective. Organized human society involves highly developed communication between individuals; the evolution of the brain and speech has led to associative and cumulative advantages which are truly unique and characteristic of human behavior. The very existence of the conceptual thought process is attributed to the faculty of verbal *communication*, which is also the vehicle for accumulation and dissemination of knowledge. Increasingly sophisticated communications thus first became the tools for collective experience retention and dissemination; individual and collective education and motivation leading to enhancement of group cohesion rapidly followed (in terms of the evolutionary time scale) as the positive differential survival value of these functions became apparent. In prehistoric times, the genetically evolved brain, together with relatively minor changes in the speech organs, essentially satisfied the information processing and communications requirements of human social behavior.

Recorded history (by definition, in point of fact) starts with the invention of writing. Functionally, this development is an *external storage device to the brain* closely coupled with the eyes and hands serving as input/output "peripherals." By this specialized contrivance, unique among all living species, humans increased by orders of magnitude the accuracy, the retentivity, and the transmissibility of information. It became possible to communicate in fairly detailed terms with distant groups and with future generations; the relative cohesion and permanence of large empires were the immediate[1] societal consequences. From a different perspective, the communication specialists, priests or scribes, became essential to orderly government; corresponding status and rewards accrued to these castes as a matter of course.

Another, perhaps just as important, revolution occurred roughly in the 15th century A.D. (in the West) when the means of writing became amenable to mass production through the invention of printing. The impact on society was spectacular and far-reaching; once again, the accuracy, the retentivity, and the transmission speed and distance of information increased by several, and in some cases by many, orders of magnitude. At the same time, the cost per unit information was also reduced to a small fraction of what it had been previously. Even the nonprofessional could reach via direct access the impressive storehouse of accumulated human knowledge. The consequences followed in rapid succession: The discovery and spreading of new cosmology, leading to distant explorations and to the rise of colonialism;[2] the Renaissance, the Reformation; and, pursuing the evolution of scientific thought, the rise of experimental sciences and applied research, eventually leading to the spread of industrial revolution. Rapid evolution of diverse political systems, indeed often taking the form of revolution, was tightly interwoven in this

---

[1] Less than a few hundred years are "immediate" in this perspective. The evolution of the imperial form of government and the routine use of written records were in all probability contemporaneous and interactive.[5]

[2] Colonies, in the modern sense of the word, can be traced back to the ancient Mediterranean cultures (Phoenicia) and were an established feature of the Greek and Roman times. Here, the emphasis is on colonial systems at the intercontinental scale.

sequence. *The intimate tie between advances in information-handling techniques and the rate of cultural, political, and social evolution is a fact well supported by historical evidence.*

In the past 25 years a new cluster of technical discoveries in the West has resulted in nothing less than the development of an *information processor auxiliary to the human brain*, namely, the electronic computer, together with the associated high-speed, reliable electromagnetic telecommunication capability. The conceptual nature of this development and its potential impacts on society are *comparable to those of the invention of writing.* For the first two decades the new powerful tools were essentially in the hands of professionals: computer scientists, software architects[3], and programmers. Less than five years ago, still another development came about, conceptually equivalent to the invention of printing: the microminiaturized electronic hardware which puts information-processing capability at the disposal of the minimally trained, nonprofessional individual.[4]

The rapid spread of electronic information processing and transmission are likely to result in a much accelerated rate of cultural evolution. The very nature of this development contains the elements of self-reinforcement; it further accelerates research and development, it promotes rapid diffusion within the national community and among nations; in short, it is a change which contains intrinsically the driving forces for further acceleration of that change. *It is apt to change rapidly and radically both the socio-political and the geopolitical climate.* The end of the dynamic phase is certainly not currently in sight.[5] Both the unusually high

3 "Software" is a specialized term in the information-processing fields; it is synonymous with hierarchy, connectivity, logic, and procedures.

4 Note in passing the compression of the time scale: It took almost 3,000 years to go from writing to printing, but less than a quarter of a century from large-scale professional computers to individual interactive equipment.

5 Titles such as "The Age of Discontinuity" and "Future Shock" bear witness to the uneasiness experienced by society facing these changes.[6,7]

return on investment in high-technology industries associated with "informatics"[6] and the rapid growth in military applications help to sustain continuing high level of research and development. Some knowledgeable observers suggest that the surface of the consumer microelectronics market has barely been scratched to date.

* * *

Many and far reaching are the consequences; it is hardly possible to even glimpse the complete picture at this time and in this place. Some of the important conclusions pertaining to our central theme can be, however, briefly summarized.

The "information environment" of the individual in modern society is rapidly becoming *more intensive, pervasive,* and *manipulative.* The message flow reaching the average individual as a citizen, a worker, or a consumer has been conservatively estimated at hundred to thousandfold the corresponding levels of as little as 50 years ago. For specialists the estimates are, of course, much higher. The rapid succession of messages precludes selective acceptance on basis of pertinency and reliability. The information flow is pervasive in the sense that a large portion of it enters physically our conscious (and even subconscious) brain, sometimes against our will, often without our awareness or consent. Only minimal imagination or interpretative effort are required by television and radio broadcast; one can even hear of new communications techniques based on the total sensory surrounding concept. A deliberate individual and nonconforming effort is necessary to shut out even partially some of the undesired messages. The use of the word "manipulative" is not intended to connote nefarious intent; it means that the production and dissemination of messages having become a mass production industry, the unit costs are now so affordable that message sources with any purpose (beneficial or harmful to the recipients) are in

6 Word coined by the French to designate the technology of modern information processing and transmission.

position to reach selectively the intended audience with little difficulty. This rich *information environment* has become an essential and growing part of our daily lives.

Scientific and technical research is now possible at rates of progress heretofore unheard of in areas previously inaccessible to human inquiry. This arises from the advent of sophisticated instruments coupled to the ability to gather and analyze information at affordable costs. The rates of advancement offer intellectual gratification to the participants and (alleged) social or commercial benefits to the sponsors. Cosmology, for instance, the science which deals with the conceptual fundamentals of the universe, has made impressive strides with advances in space sciences and theoretical physics. Paralleling Copernicus, Kepler, and Galileo, these new views are bound to affect the philosophical and religious views of many professionals and eventually those of the public at large. Molecular biology, another instance, is now at the point where it aspires to explain within a matter of years the purely chemical basis and origin of living matter—eventually of self-perpetuating individual organisms. This, of course, is still another unsettling perturbation for those who hold life mysterious and human life sacred. When the course of these endeavors threatens to synthesize, or at least manipulate, human genes, the potential impact on society becomes evident.[7] Not far behind, as a scientific endeavor of major concern to the public, is the understanding of, possible interference with, the functional processes of the human brain. Originally innocuous and beneficial, this effort was aimed at correcting dysfunctions and alleviating pain, but very soon, almost imperceptibly, the focus shifted to exploratory work in chemical and electrical stimulation or tranquilization, with the ominous overtones of behavioral control through remote reward/punishment techniques, and eventually to the centrally directed control of the individual mind. Small

---

[7] The research community, seeking to allay adverse public reaction, has imposed a voluntary moratorium on many aspects of gene-manipulation experimental research. This is a rare instance where a professional publicly expresses doubt of the advisability of furthering certain avenues of research.

wonder that the public, as an instinctive attempt at self-protection, clamors for Government control of this type of research.

***

To restrict our scrutiny of societal information flow to signals and messages deliberately generated and transmitted would be a grievous mistake. In doing so, we would also dismiss from our purview the age-old essential communication elements which, again with the help of modern technology, might offer substantial hope of mitigating the manipulative and disruptive features of the information flow which we have just discussed.

Direct, person-to-person human contacts enrich perceptions and add essential, nonverbal components to communications.[8] With the widespread availability and convenience of long-distance travel, especially since the introduction of jet-propelled commercial air transportation, direct human contacts between distant groups and vastly different cultures have become frequent and, in many cases, available as a matter of routine to large fractions of the population. It is a matter of importance that these fractions include not only the governing elites, but also (at least in the Western world) the students and other young adults whose attitudes are less constrained by the past and more concerned about the future. By encouraging direct contacts between people, by offering opportunities to examine and to explore life styles, working conditions, and the true political climate as seen by the population, Western governments can go a long way to attenuate stresses due to perceptions beclouded by distance and distorted by pejorative propaganda.

*

8 Appendix - Note B.

The implications of highly intensified and widespread information flow within and among societies have barely been explored here; some others are discussed under subsequent headings. But the thoughtful reader will already have caught a glimpse of the argument: In the presence of such a momentous change in the "information environment," a change which is so closely coupled to the essential mechanism which ties individual members of society together, a change which could not have been possibly foreseen by the framers of the Constitution, one is justified in suggesting that a critical examination of how modern society relates to the production and dissemination of information is very much in order. We should, for the same reasons, explore the international implications as well, but perhaps even higher priority should be accorded to the opportunities inherent in the more frequent personal contacts among distant populations brought about by the increasing availability of rapid means of transportation.

CHAPTER V
Submergence of Social Benefits

THE BENEFITS OFFERED BY "POST-INDUSTRIAL" SO-
CIETY TO INDIVIDUAL MEMBERS BECOME SUBMERGED
BY CONFUSION, COMPLEXITY, AND CONFLICT

Condensed in its starkest simplicity, the cohesion of a social
unit is determined by the benefits offered to its members in
exchange for the burdens imposed upon them by society. In
a stable and cohesive group, the benefits or rewards, *as
perceived by the recipients,* must outweigh, or at least
reasonably balance, the constraints and servitudes to which
an individual is subjected by the mere fact of being part of
the group. Rewards are perceived in general as equitable
when they are in some reasonable proportion to the social
usefulness of the recipients; the distribution of burdens is
still subject to many conflicting opinions. Both rewards and
burdens include the tangible material aspects, but a strong
case may be made for the psychological elements being at
least equally important in forming the image conceived by
individuals regarding their relationship to society (their
"self-image").

Following World War II, the Western democracies (with the
notable exception of Japan under the MacArthur Constitution) have all experienced a conspicuous negative trend in
the commitment of their people to the prevailing systems of
government. This has become mostly apparent since the
mid-1950's as widespread disenchantment with the leaders,
apathy and cynicism for public issues, and loss of faith in
the longer term, successful future of the Western-style
democratic institutions. Active hostility and desire for

radical or even violent changes (culminating in sporadic
terrorism) became fashionable in the late 1960's. Since the
general availability of material benefits unquestionably
increased during the same period[1], one may be justified to
assert that all these "democratic" societies have, in some
important ways, failed to provide the psychological rewards
or stimuli (equally perceived by the people as important
"benefits") which contribute so essentially to their cohesion
and stability. This is the assertion to be explored now in
somewhat more detail.

*

A general atmosphere of insecurity and discomfort
permeates the thinking of a large proportion of middle- and
working-class populations. The political theory holds that
government is "...for the people" and that "the just powers
of the government are derived from the consent of the
governed," but the individual citizen increasingly perceives
the government as remote and indifferent, when not downright obtrusive or hostile. Issues directly affecting his life
appear hopelessly complex; he has long ago abandoned any
serious commitment to understand or to foresee the consequences of his choices; his chances of exercising deliberate
control over his own future are, in general, increasingly
slender. Many issues, even those which may affect vitally
the relationships of the individual to society, are now
certainly complex beyond the understanding of untutored
laymen; they are quite often beyond the capabilities of most
experts as well, so that professional guidance is hard to
come by. The primary cause of the increased complexity is
no doubt found in the vastly broadened interactions within
each group and with (socially or geographically) more
distant groups having conflicting goals, diverging value judgments, and incompatible beliefs. Contrasting with the
recent past, there is quantitatively much increased (if not
always applicable) analytical capability on hand and the
rapid accumulation of (not always pertinent or accessible)

---

[1] A significant proportion of violent protesters and even terrorists have no
consistent background of harsh material deprivations; well to the contrary,
many of them come from middle-class families.

Copyright 1979 by Herblock in The Washington Post

empirical data base further reveals the complexity and interdependence of many public issues.[2] Attractive solutions or compromises would call on combined expertise in several scientific or technical disciplines, melded with wisdom, political science, and common sense value judgments. Unfortunately, the analytical or empirical foundations upon which sound decisions can be made are especially deficient in social sciences, and large-scale experimentation is fraught with unpleasant, or even irreversible, consequences.[3] The traditional professional classes (lawyers, doctors, educators, scientists, and religious leaders) have by and large ignored the challenge of these newly emerging societal problems. Some of them are actively engaged in rear-guard action to protect their professional prerogatives; the public retaliates by increasing the incidence of professional malpractice actions. Some others use the still extant authority associated with learning to claim competency in areas outside of their expertise and thus voice their own prejudices (often just as intense as those of the laymen). Very few think in terms of the possible broadening or combination of their professional background and talents to attack problems (or perhaps nonproblems) that contribute to the almost permanent public crisis atmosphere. The citizen is bombarded by the daily bad news, feels inadequate to evolve his own reactions, cannot find impartially competent guidance; his reaction is increasingly passive if not downright hostile. He is apt to vote on emotional grounds, if he votes at all; his participation is at the best perfunctory, and his sense of true participation has vanished long ago.

2 Appendix - Note C.

3 Erroneous approaches, by now publicly criticized and retracted by the very authorities instrumental in their earlier advocacy, can be found in recent history. Desegregation of schools to improve Black education and permissive child rearing to improve ultimate chances of social adjustment are signal examples.

(Note added in proof) The notorious snail darters, protected against the destruction of their habitat by the full authority of the U.S. Supreme Court, have been discovered as living in a happy state of obliviousness in several nearby rivers by the same expert biologists who originally and forcefully insisted on their state of being endangered.[8]

*

The livelihood and the secure place within society of many people are threatened by the rapidly changing skill requirements inherent in the organization of modern production and service activities.   Automation in manufacturing and process industries, in office and clerical-type occupations, in agriculture and extractive industries is driven by technological advances and by competitive pressures.   It eliminates or downgrades many jobs which only recently were described as "highly skilled"; those requiring medium or low skills were subjected to erosion long ago.   The displacement of workers occurs at a faster pace than any reasonable retraining program could accommodate.   It is no longer true that displaced workers, or at least the most talented among them, rapidly find employment in the production of automated machinery or in the implementation of automated processes which have led to the problem in the first place.   At the other end of the skill spectrum, minimum-wage laws have contributed to the elimination of many marginal jobs, often the only ones available to those at the unskilled entry level.   Unions fight a rear-guard action (e.g., in railroads and printing industries) but the members know all too well that, in the long run, their relatively well-paying jobs will be eliminated and replaced by others to which they can lay no claim by virtue of experience and seniority.   The relatively rapid rate of change in skill requirements is driven by the classical forces of profit motive and competition abetted by technology; the impact on workers is only slightly tempered by public efforts to retrain those who are individually threatened by the process. It is possible, although quite difficult to prove, that the skill requirement spectrum imposed by the direction of the post-industrial economic investment is in the process of shifting beyond the capability spectrum of the composite labor force which can be reasonably anticipated for the future. [4] If this should occur, adverse social consequences will inexorably follow: insecurity, boredom, demotivation, and loss of self-respect in the work force,

4 Appendix - Note D.

increasing reliance on public efforts to "create" jobs and to ensure nonetheless (in some mysterious but unspecified ways) "decent" standards of living.

*

Society's decisions are distorted by the rapid emergence of numerous broadbased pressure groups not properly subjected in time to the traditional checks and balances. This, to a large extent, is a consequence of the availability of mass communications and data processing techniques which allow effective joint action by individuals sharing some special interest in common. Some of these groups, such as the poor, the aged, and racial minorities, have existed for a long time but have recently achieved marked prominence in the political process. Some others are of more recent vintage, such as the environmentalists, the advocates of national health service, zero population growth, pro-(or anti-) abortion legislation, moratorium on nuclear power development, and of other causes innumerable. In the past, the slow accretion of any constituency was accompanied by reasonably contemporaneous development of countervailing "checks and balances."[5] The new aspect of the situation is that (1) the growth to the level of strong political influence is now measured in years, rather than in decades, thus inroads on the nation's attitudes and distribution of resources can occur *before* the countervailing checks and balances could become operative; and (2) the financial interests of the beneficiaries leads to the coalescence of congressional groups, lobbies and civil service constituencies which, together, make effectual opposition hazardous to any politician who might be concerned with the interests of society taken as a whole. The underprivileged, the jobless, the minorities of all descriptions, the supporters of many plausible-sounding causes can thus combine and exercise considerable political power, and not necessarily to the lasting benefit of society.

[5] E.g., the monopoly of railroads was checked by the establishment of the ICC; big business was restrained by the growth of antitrust legislation; both big labor and big business found their respective powers circumscribed by the Wagner Act; at a later date labor found its power curtailed by many state-level antiracketeering and right-to-work statutes.

Causes are not meritorious just by virtue of the fact of being supported by vocal minorities; the legislatures are clearly inappropriate to adjudicate the objective merits of a given proposition when the support given to causes with strong emotional appeal may mean the difference between reelection and defeat for the legislators. It has also come as a shocking revelation to those attuned to liberal viewpoints that the poor, the sick, and the victims of discrimination, and more generally those who call upon our compassionate generosity (and, not incidentally, those who assiduously pretend to cater to their needs) are not necessarily virtuous or public spirited. In fact, a large cost fraction of social benefit programs is attributable to the poor and the afflicted being just as prone to abusing the advantages of the system as those who are more fortunate. In a similar vein, while the young (and even those not so young) "activists" of all sorts often pose challenging questions, it cannot be said that they, as a group, possess the singular combination of wisdom and foresight that would make their opinions infallible or their recommendations incontrovertible. The net result of many groups being able to modify the allocation of societal resources is over-compensation and distortion of the responses, which, of course, waste the nation's energies and substance, and create additional confusion and insecurity in the public mind. The responses to even the legitimate needs become delayed or distorted as well. [6] For the individual, the resulting confusion, the insecurity, and the conflict among rival claims are translated into a widespread feeling that society is being manipulated in a direction contrary to his real interests, that his taxes unfairly encourage idleness and abuse among those who are less diligent or less provident.

The natural reaction of individuals is to shift their concern to narrow, immediate issues, to forget the longer term, broader considerations, and to take advantage of the "system" whenever possible. All this, of course, is a far cry from attitudes leading to responsible citizenship.

---

[6] Examples are plentiful: The recently discovered energy crisis, our lagging capital facility investment, our dependency on foreign raw materials, etc.

*

The Federal Government's intervention in the nation's economic activity is perceived as ambiguous, if not plainly detrimental. Many instances could be discussed; the anti-smoking campaign pursued while subsidizing the tobacco farmers, fighting inflation attributed in part to the high food prices while at the same time restricting meat imports, being prime examples. Perhaps the governmental actions associated with the regulation of consumer credit deserve discussion as embodying the most glaring mutually contra-dictory elements. Governments at all levels object to high interest rates as a matter of principle, since they "add to the burdens of those who are the least able to carry them." Several states, in fact, impose "usury" laws to limit the maximum interest which can be charged. [7]

When inflationary pressures mount, the Federal Government helps, nonetheless, to increase the prevailing interest rates through policies pursued by the Federal Reserve system. This, of course, further exacerbates the inflationary pres-sures, and (not so incidentally) increases the profits of lending institutions. For some mysterious reason, excess profit tax on banks has not been forcefully proposed in recent decades; in fact, it is hardly mentioned at all. Even worse, the Government condones, when not explicitly encourages, consumer credit growth (as contrasted to investment credit, such as housing mortgages, private capital facility, bonds, etc.) by generously making interest payments deductible from taxable income. When buying interest is lagging, the Government joins the campaign to encourage spending [8] in the interest of maintaining the economy in high gear. The explicit governmental actions

[7] Interestingly, the very small loans are sometimes exempt from interest ceilings, either explicitly or through various subterfuges, even though the annual percentage rate must be disclosed.

[8] The "YOU AUTO BUY NOW" campaign of 1967 could be cited as an instance.

(Note added in proof) Since January 1980 the Federal Reserve Board has attempted to clamp down on credit by allowing interest rates to reach levels clearly considered criminal since biblical times and, in several States, until as recently as 1976.

added to the vigorous advertisement campaign by the producers of consumer goods result in a staggering growth rate
of consumer credit. The economic impact is most
burdensome[9], but we are mostly concerned here with the
psychological aspects of this trend. To stimulate consumer
interest, the terms of repayment are thoughtfully extended
well beyond the period of ecstasy stemming from the new
acquisition; in fact, they often exceed the useful life of the
product. Thus, a growing number of wage earners sense that
an increasing fraction of their discretionary income (i.e.,
that which should have been originally available to finance
enjoyment and investment) is in fact paying for the sequels
of past enjoyment, something which is almost universally
resented and sometimes resisted. At the national level, a
congeneric uneasiness pervades the thinking of those who
are paid to engage in policy-making; the substance of
tomorrow's earnings is being spent, not on productive
investment, but on today's or even yesterday's, superfluities
artfully disguised as concomitants of decent standard of
living. The thought that this cannot continue forever is
logically most compelling, but to bring any fundamental
remedy to the situation is for all practical purposes inconceivable. Here then is a situation which citizens and policy-
makers equally perceive as undesirable, and clearly leading
to disaster if allowed to continue, but under the guise of
respect for the consumer's freedom of choice, no real
corrective actions are taken.[10] The individual citizen,
seeing his net discretionary income (that which remains
after the installment loan payments) shrinking even further
by the relentless growth in inflation and taxes, eventually
concludes that there is no tangible relationship between his
productive efforts and the incremental enjoyment he derives
from his income. This may well be the operational definition of slavery for modern times. The results, when carried
to the extreme, are growing degrees of indifference, aliena-
tion--and possibly search for a convenient scapegoat: the

[9] Appendix - Note E.

[10] Ironically, the same citizen is not trusted to make the relatively simple
judgments which would render many activities of the Consumer Protection Agency superfluous.

"system" or the "establishment." Here again, we find that the resulting trend is definitely not in the direction of increased societal stability.

*

Even though the objective statistics may not bear out this impression, society appears to fail in one of its most widely accepted roles: that of protecting the physical security of its law-abiding citizens. The incidence of crime, violent protest, political assassinations, random terrorism may be statistically low for any one individual, but aided by the distorted focus and the cumulative magnification of mass media, to many the threat appears as being real and immediate.

*

Most governments of the Western democratic regimes have consistently failed to provide the moral tone of leadership expected by the populations. No doubt people are far too sophisticated (or disabused by repeated exposure) to believe that their leaders are superior human beings, but there is, nonetheless, a subconscious expectation to have those who, after all, shape our individual destinies live their lives and conduct their public behavior according to commonly accepted standards of morality. Those who recall, to use once again the U.S. as an example, the sorry litany of Vietnam, Watergate, Korean payoffs; the individual capers of Congressmen Mills, Hayes, Floyd, Diggs, and of many others; the publicized "misdeeds" of the CIA and of the FBI just might gather the impression that such expectation is largely unwarranted.[11] The recent history of Great Britain, France, Germany, Japan, and Italy hardly offers a more encouraging picture.

---

[11] (Note added in proof) The recent sensational instances where the FBI went to extreme efforts to secure videotaped evidence of misdeeds among Senators and Congressmen elicited much public interest but hardly any surprise.

*

In summary, an increasing number of instances can be shown where the issues confronting the individual are complex, often beyond his understanding, where the multiple arguments are confusing, where several objectives are conflicting, and where dispassionate professional guidance is hard to come by. This newly evolving trend affects the nation's internal environment inasmuch that the many material and psychological benefits offered by modern society are taken for granted and are increasingly submerged in the minds of many under the insecurity engendered by this confusion, complexity, and apparent conflicts. The emotional attachment of individuals, their enthusiasm for the participatory political process are being eroded; calls for sacrifices in the interest of the public good are apt to find a cynical, rather than a receptive, audience. Societies with such characteristics are possibly wide open to manipulations by clever opponents who can intensify the specter of external threats, skillfully modulated to engender alternatively hysteria and complacency. A society exhibiting such trends is becoming vulnerable indeed.

CHAPTER VI
Ecological Constraints

THE ECOLOGICAL CONSTRAINTS ON NATIONAL AND GLOBAL GROWTH ARE BEING RECOGNIZED AND MORE GENERALLY ACCEPTED

Man is part of nature; he has been vitally concerned with the ecology well before the word was coined or became sacrosanct. His concerns have, since times immemorial, focused on two aspects: the "despoiling" or "exhaustion" of the resources offered by his environment and the "pollution" or deleterious change in the environment, stemming from what is often being described in modern times as the "irreversible fragility of the ecology."[1] Both concerns are justified and serious but must be placed in context. What is meant by exhaustion is that a given resource is no longer accessible to exploitation with the currently available techniques at essentially the prevailing economic cost. While shortfalls in available resources were a quasi-permanent feature of human history, they were usually alleviated by migrations, wars, changes in technology or in consumption habits.[2] Only in modern times have people begun to express concern at the global level; whether such global concerns are justified can in many cases not be fully decided at this

---

[1] The word *fragility* is not quite appropriate. What is usually meant is that modifications beyond relatively narrow limits of apparently noncritical elements in a given ecological unit cause some of the results to become cumulative, self-reinforcing and irreversible.

[2] Changes in reproductive behavior and in mortality are characteristic responses in the animal kingdom as well as in some human groups.

stage of our knowledge. The irreversible modification of the local environment (or to use the obviously pejorative term "pollution") is also a phenomenon recognized as being part of human evolution since prehistoric times; it is also relevant to animal species. The immediate cause of danger to primitive human groups, the proliferation of bacterial infections and other parasites in the vicinity of garbage mounds, was usually remedied by exmigration.

The newly increased recognition of the dangers of resource exhaustion and pollution can be traced back to many confluent trends. Increased industrial activity of the past 30 years and the more intensive agricultural development have brought about growing levels of chemical pollution of the environment. The efforts aimed at eradication of disease vectors (mosquitoes and rodents) have further added to man's nondegradable chemical arsenal. The impact of pesticides and insecticides has been eloquently described[9] even though some of the early statements and conclusions have been found, upon further examination, one-sided and somewhat exaggerated. The increase in off-shore oil exploration and of seaborne oil transportation, coupled with the economic pressures leading to the construction of ever-larger supertankers, raises the frightening specter of oil pollution of the seashore as well as of the broad open ocean. The horror stories of the Santa Barbara oil leak, the accidents of "Torrey Canyon," "Argus Merchant,"and of many others are vividly in the mind of environmentalists and their audience.[3] The damage to fresh-water lakes and rivers by industrial and municipal effluents and the unintentional but well-publicized side effects of major insect control programs all have contributed to create, by overreaction characteristic of postindustrial societies, a widespread and deep-felt concern about nature, wildlife, and wilderness areas to be preserved in their state of pristine purity.

The emotional revulsion against the fallout from nuclear weapon testing among populations improperly prepared in advance (and subjected to strident, if distorted, propaganda on the subject) has prompted reluctance to accept even the minutest potential of radioactivity levels. Even the problem of radioactive waste disposal is apt to create an irrational scare completely out of proportion to its proven or predictable biological hazards. [4]

As the pharmaceutical industry attacks more and more complex biological problems and develops more potent medications with difficult-to-predict side effects, catastrophies are bound to happen. The thalidomide uproar in 1962 was amply justified and it is indeed fortunate that vigilant doctors (in Australia) established the disastrous cause-to-effect relationship at a relatively early stage of clinical applications. In the minds of many the dangers of potent pharmaceuticals, added to the carcinogenic and teratogenic dangers due to occupational exposure to toxic products, should create a prohibitive atmosphere toward any recognizable artificial component of the chemical environment, even at exposure levels detectable only by the most exquisitely sensitive instruments.

The suspicion against potentially dangerous substances goes much beyond the domain of pharmaceuticals. The increasing evidence linking smoking to lung cancer, the recurrent actions by the FDA (almost always justified) against additives, hormones, colorants, sweeteners, and the proliferation of processed convenience foods has by now established a public climate in which any new synthetic substance is viewed with reluctance if not alarm and even the old time-tested "natural" food ingredients are questioned.

In the mid-1960's, a relatively small group of concerned scientists (Club of Rome), supported by a competent group with expertise in computer model development, proceeded to describe the large-scale future behavior of the global ecology, in particular as it affects the human population.

3 Characteristically, a new breed of lawyers, specialized in "environmental litigation" stands ready to help those at risk (fishermen, beach property owners, and resort operators) to collect handsome damage settlements, even when substantial damage in the long run is hard to prove in court.

4 Ironic as this may sound, the total radiation hazard in the U.S. to humans from the currently planned coal utilization levels in the 1985-2000 year period exceeds that of all the radiation sources possibly associated with the original (1000 units of 1000 MW each) nuclear powerplant building program, including waste disposal. [10]

Based on models which are by now judged rudimentary by most investigators, they concluded that, irrespective of the choice among scenarios, the world is facing violent near-term instabilities and upheavals. In particular, their model suggests that the advanced industrialized nations will suffer a far-reaching and possibly discontinuous degradation in their standard of living—in short, a sequence of catastrophies.[11] The models are by now recognized as having failed to account for the remarkable adaptiveness of many human activities; also, many conclusions were based on extrapolation of time sequences which by no stretch of imagination can be considered statistically stationary. Those concerned have, however, found a convenient quasi-philosophical support for their beliefs: the finiteness of the Earth's resources has become a popular argument in favor of limiting population growth and restricting resource consumption.[12]

The public concern about pollution of air, water, and soil has been recently translated within the U.S. Federal Government into the formation (1971) of the Environmental Protection Agency (EPA) with its counterparts at the state government level and with the spread of similar control agencies to most Western nations. Coupled with much improved analytical techniques, capable of detecting pollutants in proportions of the order of one part per billion, the EPA and associated legislation have paved the way to the mandatory internalization of environmental "cleanup" costs. In many instances, the goals set by legislation are exceeding the environmental quality levels set by natural pollution.[5] The economic cost of marginal improvements in environmental quality diverts capital from productivity-oriented investments.

Increased scarcity of resources and simultaneous attempts to protect the environment result generally in clearly perceptible increase of the societal overhead costs, unless new and highly effective technical solutions are introduced.

---

[5] An apocryphal, but quite believable, story holds that a broad expanse of wilderness in the Oregon Cascade Mountains has been found in substantial violation of the EPA's clean air standards. The offenders, the tall Douglas firs of that part of the country, seem to exude on warm days a mixture of aromatic compounds clearly and explicitly prohibited by law...

In practical terms, this means that the cost of materials, goods, and services is increased in relation to the cost directly associated with their production. Whether this is or is not desirable from any longer term viewpoint is not easy to decide, since many of the associated value judgments are perforce subjective. It is not obvious, for instance, where the compromise should be drawn between the availability of cheap lumber (i.e., lower housing costs, affordable urban development, and decent quality-of-life for the disadvantaged) versus the enjoyment derived by future generations from broad expanses of wilderness, preserved in their pristine condition, physically accessible but sternly and forever forbidden to the run-of-the-mill citizen. What is to be expected though, and clearly shown by experience to date, is that as we increase the emphasis on the protection and restitution of the environment; as we endeavor to clean up our surrounding air, water, and soil of any possible trace of human exudates, the cost of human biological or economic activity will substantially increase. This is one, and certainly not a negligible, component of the inflationary pressures which have arisen among the industrialized nations in the last decade or so.[6]

While the results of concern for the environment add up predictably to increasing society's overhead costs, other consequences may well be at a much larger scale and are often unsuspected at the time when ecologically significant moves are being undertaken. As an illustration, in the early '70's the attempts at the eradication of the tsetse fly in sub-Saharan Africa have made considerable progress; indeed, for

---

[6] One typical example is offered by the unintended effects of the EPA-mandated automobile pollution control. As air quality standards were set in 1968, they first resulted in lower efficiency in oil refinery operations by mandating increased production of unleaded gas. This also has required massive additional refinery construction, with further repercussions on gasoline costs, when in actual fact, gasoline consumption should have remained stationary or even decreased in view of the increasing crude oil costs and the growing emphasis on energy conservation. In a connected development the hasty attempts by the automobile manufacturers to introduce catalytic converters has decreased operating fuel efficiency and also increased the cost of acquisition and ownership of automobiles. As a result, fuel prices and automobile prices have increased in concert to bring home to the average consumer the very tangible dollar cost of automobile pollution control.

the first time since well before the biblical era, cattle could be efficiently bred and sustained in tropical Africa, and more specifically in the Sahel region. The economic benefits to the individual cattle owner in terms of labor, food, and "capital gains" are of such obvious magnitude that herds have rapidly multiplied until the scant grazing resources of many marginally desert areas have become overburdened. This being the case, the slightest adverse climatic fluctuation results in irreversible spread of the desert into areas theretofore used as pasture.   This "desertification" is exactly what in fact happened in the 1972-1973 period.[13] The inroads of desert with the possibility of permanent climatic changes due to the elimination of grassland has become a definite threat to the region.

Still another example should be cited to illustrate unsuspected and adverse consequences of even those undertakings which at the onset offer the most bountiful promises. The Aswan Dam on the Upper Nile was undertaken in the late 1950's with massive Soviet financing and participation.[7] It was planned and advertised to provide abundant irrigation water ("to make the desert bloom") and electrical power sufficient for a most ambitious industrialization program.   The reality, however, turned out to be quite different.

Artificial irrigation in an area of high evaporation, in contrast to the yearly flooding of the fertile strips directly adjacent to the Nile, results in salt saturation of the otherwise fertile soil. The natural silt of the river is no longer available to rejuvenate the traditionally cultivated alluvial areas so, ironically, an important fraction of the power generated by the dam is being used to manufacture artificial fertilizers.   The net result on the agricultural

---

[7] The U.S., under the guidance of the late John Foster Dulles, rejected the financing of the projects for reasons (needless to say) having no relation to ecological concerns. Among the political consequences (had someone been wise enough to file a political impact statement at that time) many would mention the 1956 Arab-Israeli war, the increased Soviet presence in the Mediterranean, the intensification of the Algerian "national liberation" war, the overthrow of governments sympathetic to the West in Iraq (1958) and Libya (1960).

output of the nation is a matter of controversy but, at any rate, far below the early, optimistic predictions. On the other hand, the retention of silt above the dam in Lake Nasser promises future upkeep problems of no inconsiderable magnitude. More importantly, it has just about wiped out the Mediterranean sardine fishing industry, which was essentially dependent on the influx of mineral laden silt to ensure the abundance of fish harvest each year. Still under the heading of unsuspected consequences, we may consider the much increased incidence of bilharziasis, a debilitating tropical disease spread by snails[8] that thrive in irrigation canals. Bilharziasis has long been widespread in the Delta but, since the High Dam began providing water year round, the disease is on the increase in Upper Egypt.[14] The list of unexpected side effects could probably go on.

In the two instances just shown, a valid argument can be made to the effect that a competent "environmental impact statement," prepared after serious consideration of all the known factors prior to commitment, could have greatly mitigated the adverse, unintended side effects. It is also true that the original aim of the undertakings had no explicit relationship to the protection of the environment. By way of contrast, in some of the currently raging controversies over the future of nuclear power, the clearly proclaimed objective (in this case, of the environmentalists and of the absolute opponents of nuclear power joined by advocates of no-growth, of nuclear weapon ban, and others) is the preservation and protection of the wholesome aspects of the human environment. Their position is that nuclear power plants are *unnecessary* (we can do with lesser energy requirements, if only people would curtail consumption); that they are *unsafe* (behold the dire consequences of the Three Mile Island type accidents); that they are *poisoning the environment* through radiation, waste and thermal effluents. All these objections, of course, can be (and have been, at some length) answered by knowledgeable people;[15,16] the general availability of energy is the primary factor underlying the growth of the economy, the

---

[8] Bulinus truncatus (North Africa)

only source of real wealth to be shared by a growing population; the actual safety record and projections of the nuclear power industry are far better than those of any other undertaking of comparable size and complexity in human history; the prompt and delayed radiation hazard can be reduced to well below the natural background level by known and intelligently applied safeguards; as to the thermal "pollution," in some rivers and estuaries warm water has been convincingly shown to attract and nurture aquatic life in unexpected abundance and variety.

The real issue, however, is elsewhere. While the self-anointed champions of the environment pretend (and possibly believe) to have carefully evaluated all the consequences of their position, in fact they circumscribe their concern to some selected and unilaterally perceived aspects of the environment. If the need for adequate energy supply is conceded, the radiation and safety hazards associated with alternative energy sources must be considered. Coal has been a hazardous and unhealthy undertaking and is expected to remain so, mining regulations notwithstanding. Surface mining of coal also raises some difficult questions about restitution of *that* environment to its original condition. On a larger scale, consistent reliance on fossil fuels is apt in the long run to increase the atmospheric carbon dioxide content with the concomitant (still controversial) consequences in regard to the global climate.[17] If, on the other hand, the growth of our economy is to be constrained by severe curtailing of the energy supply, the advocates of no-growth are notably silent about the deleterious human environment that is likely to be perpetuated under such conditions, because of our reduced ability to reinvest capital surplus in undertakings such as water and waste management, habitat, transportation, utilities, etc. It is not possible, in our present state of knowledge, to answer many related questions; many of the most important questions have perhaps not even been formulated. But it is quite safe to predict that none of the current extreme positions will, in the long run, provide anything close to the "right" answer.

*

This last illustration is particularly instructive because it portrays vividly the conflict between the broad, longer term ecological concerns on one hand and immediate economic cost including repercussions on the rate of economic growth and on the level of productive (i.e., wealth-producing) activity. Giving way to environmental protection considerations causes our overall societal activity to become more costly. It increases the government's regulatory activity which in turn still adds to the nonproductive sector of the economy; it also further discourages individual initiative and creativity and places increased constraints upon resource development and subsequent exploitation.

At the domestic level, taking once again the United States as an example, all this results in slower economic growth, probably to the point that it will soon fail to match the rate of population growth. This, then, eventually leads to less available discretionary surplus for continuing high level of capital investment. If the political stability of the country is to be preserved (with democratic systems typically promising always more to those who have less), this means further wealth redistribution in favor of the "under-privileged." Whether or not these are characteristics of a vigorous, dynamic nation is left to the reader's appreciation.

The same argument could be presented in a perspective transcending national frontiers. Most of the important environmental problems cannot be constrained by arbitrary political boundaries. Thus, the pollution of the atmosphere, rivers, and of the ocean, or the implication of nuclear weapon testing, are obviously matters of international concern. The trend toward increased emphasis on preservation of the environment is a theme seen by the have-not nations as a luxury. Their lands are less impacted by current pollution levels and many of the less-developed countries most urgently need development and jobs. Their only sources of wealth are their mineral and agricultural resources, in addition to (mostly unskilled) labor. It is difficult to cause these countries to see pollution in the same perspective as that of the inhabitants of wealthy cities whose consumer demands have to be artificially stimulated by intensive advertisement.

Different perceptions of ecological concerns among the industrialized nations lead to new competitive elements. Thus, in the recent past it became clear that the French and the Germans, for instance, do not have the same perception of the dangers of nuclear radiation or of weapon proliferation as that which forms the basis of current U.S. policy. For all practical purposes this may, although this is far from being proven, have given to Western Europe a substantial edge in future commercial nuclear reactor development.[18] In the same vein, the British and the French have not seen fit to place the same limitations on nuclear fuel reprocessing as those imposed by the United States, once again resulting in their significantly improved competitive posture.[19] For all practical purposes, foreign sales of U.S. nuclear power plants have stopped since formulation of the relevant policy by the Carter Administration.

The overall impact on societal stability discussed in the domestic perspective can also be extended to the global level. If the industrialized nations are constrained to a much slower growth by environmental, resource, and political considerations, less capital surplus will be generated for reinvestment in infrastructure and productive capacity development. This in turn will slow down the aspirations of the less developed nations and will raise additional demands for global wealth redistribution. This is even more objectionable from the domestic political viewpoint than the redistribution of wealth inside of national boundaries.

*

We have examined a relatively newly arisen concern related to ecology and the environment owing to many confluent causes. Basically, this set of concerns increases the modern society's overhead costs and decreases the immediately available discretionary wealth resulting from productive economic activity. Presumably, such wealth is displaced into the direction of long-term "investments" in national health, well-being, esthetic enjoyment, and perhaps many other intangibles, but often such ultimate benefits tend to become controversial as soon as they cease to remain unknown.

Approaches under the guise of solutions are urged without the full understanding of consequences and possible impacts and often they result in general restrictions on wealth-productive activity; the slow-down of growth threatens the basis for political stability of democratic governments. The recognition and gradual acceptance of the ecology as a limiting factor of national and global growth rates should be considered as a significant new element in our environment; the adaptive response of our societies has yet to evolve.

CHAPTER VII
Political and Economic Instability at the Global Level

THE INDUSTRIALIZED NATIONS HAVE LOST (OR RELINQUISHED) THE POLITICAL AND ECONOMIC DOMINATION OF THE WORLD. THEIR ATTEMPTS AT REMAINING PREPONDERANT PRODUCERS OF WEALTH AND PRIVILEGED CONSUMERS OF RESOURCES ARE INEFFECTIVE AND INCREASINGLY THREATENED.

Taking a somewhat somber view of human history, it may be characterized as a multiple sequence of violent and often mortal quarrels among breeding groups, tempered only occasionally by short periods of cooperation.

While in the last quarter of the twentieth century the objective assessment of the long-term interest of mankind should mandate peaceful cooperation at the global level, many differences in local conditions, exacerbated by the much increased and often distorted information flow, persist in sharpening the subjective perceptions of near-term conflicts. Differences are indeed fundamental and far reaching; they include the imperatives of geography, climate, soil, and mineral resources; they are often reinforced by cultural, religious, racial, and demographic factors. More significantly, in modern times they are further compounded by the cumulative effects of capital investment in education, productive discipline, and facilities.[1]

---

1 In contrast to animal species, humans considered from the global vantage point suffer from a particular affliction owing to their elaborate communication technique, the articulated language. Biologically we are a single

*

It is in this world, then, so perfused with sources and residuals of conflicts and quarrels, that the Western nations must evolve their foreign policies in support of their national interests. Admittedly this is not an easy task, but for all practical purposes the alternative can not be seriously considered. Modern democracies, as discussed earlier, attempt to maintain the stability and cohesion of their societies by holding out promises to the voting majority, which quite understandably (paraphrasing Samuel Gompers' celebrated dictum) simply want "more." Growth in real terms is the only known mechanism whereby reasonable fulfillment can be offered against such promises without stresses and dislocations in the existing political and social systems. Growth in turn can be sustained only by having raw materials and energy sources accessible at reasonable cost. Once the national resources are exploited at, or close to, the limits of economic accessibility, this for all practical purposes means interaction with regions where this limit has not yet been reached. In short, prosperous economies in industrialized nations are strictly dependent on continuing beneficial interaction with the resource-rich Third World. The level of industrial production, if fueled by raw materials, energy, capital investment, and the relentless pressure of technological innovation, soon exceeds the consumption rate within national boundaries, even when over-stimulated by advertisement and consumer credit. So, the Third World is also invited to absorb our production surplus if only it could pay for it, preferably out of its own earnings.

1 (Footnote continued)
    species capable of interbreeding, except for cultural barriers. Pursuant to geographical dispersion, individual groups within an animal species soon acquire (in terms of the evolutionary time scale) different characteristics and occupy slightly different ecological niches, so that, if brought back into contact, direct competition is much attenuated. By way of contrast, the human species has managed the simultaneous feats of having erected cultural barriers to interbreeding and language barriers to understanding, while having also invented the technical means for interacting and competing (communicating, trading, and fighting) at global distances. No wonder that we seem to have a problem.

There is a fundamental source of conflict here. Our producers would like to have low costs for raw material, energy, and unskilled labor (when required). They also want high prices for their finished products, the capital goods and the services they are in position to offer. Our trading partners in the Third World, not surprisingly, insist exactly on the opposite. In addition to cheap credit and guarantees against commodity market fluctuations, they strive for high prices for their raw materials, and good wages for their labor; they also want reasonably low-cost access to production tools, industrial products, and the Western consumer goods. In recent time other demands, such as underwriting of basically uneconomical pursuit of social objectives, have also become prevalent[2] if not popular.

In past times (specifically referring to the period between roughly 1650 and 1910 or perhaps 1950) this fundamental conflict of interest and viewpoints was resolved or at least subdued by a combination of several peaceful mechanisms, vigorously supported by some others, perhaps not so peaceful. In earlier periods of what eventually evolved into the "colonial/mercantile system" trade was facilitated by immense economic differences between the newly discovered regions and the home base of European explorers. Typically, and for a short time, gold, spices, and slaves with high trade value at home or in other regions could be obtained in exchange for trade beads, mirrors, bells, calico, and somewhat later for tools, utensils, and intoxicants.[3] The

---

[2] Clauses which explicitly relate to social objectives of resource supplier nations are often only marginally pertinent to a specific transaction. They have become quite frequent in the last decade or so, to the extent that foreign customers almost accept them as a matter of course.[20] For instance, the Indonesian government will insist on training programs for future Indonesian managers, with guaranteed majority of management jobs on joint resource development projects. But a country is not necessarily underdeveloped in order to resort to this "social strategy." In recent years Canada, for example, has tied major aerospace product purchases to an economically unjustified, but socially desirable, program aimed at bringing high technology manufacturing activity into its remote Northwestern regions.

[3] Tobacco perhaps was an interesting example of the reverse flow in this regard.

profits in this type of relationship were so enormous, owing to the very real economic and cultural gradients,[4] that little or no coercion was necessary.   In point of fact, the beneficiaries' concern on both sides was mostly to ensure monopoly, i.e., to prevent competitors from gaining access to the same lucrative trade channels. Partly stemming from genuine commitment, partly as a deliberate attempt at continuing subjugation, the future colonial powers had exported tenets of their respective religious beliefs and social organization.

Somewhat later we can trace the fortuitous or deliberate beginning of what today would be described as "technology transfer," mostly in support of more intensive agricultural and mineral resource developments.   The organization (a mild word for enslavement) of the local labor force was vigorously pursued with the connivance, or better the explicit support, of the local rulers.   The transfer of weapons of military technology and organization enhanced alliances and friendships, while weakening or destroying those who attempted to resist additional "Western" encroachment.   Not so incidentally, the arms trade became an additional and important source of profits.[5] By the time the colonial status was formally established and had acquired legal standing within and among colonial powers (roughly around 1700), large-scale capital investments mostly aimed at infrastructure development were under-taken by both public and private interests.   Many of these projects were supported even after the colonial regime had disappeared,[6] provided that the economic returns remained attractive and governmental protection of the investor's interest against the vagaries of local politics remained effective.   Here again, we find similarities to present day preconditions for sound overseas private investment.   Still

---

[4]   I suggest this term borrowed from mathematics to designate differences proportional to geographical or cultural distances.

[5]   While intended to describe the past, these words have been chosen to be applicable to contemporary conditions as well.

[6]   As a for instance, in Latin America the British, French, and Belgians developed most of the mines and the railroads in the late 1800's.

further reinforcement of the peaceful, stable aspects of the colonial system was obtained in some specific instances by social cooptation between the expatriates of the mother country and the native aristocratic or intellectual elites.

Many are the combinations and variations on these themes but military supremacy of the colonial powers was still ultimately the basis of stability in the colonial regimes. It also was, while it lasted, the guarantor of peaceful evolution by protecting the colonies against regional upheavals and by providing effective deterrence to unrest or recalcitrance against prevailing political, social, or economic conditions. Whenever local conditions have stymied the assertion of military superiority (e.g., Afghanistan or Ethopia), or where the potential rewards have failed to justify appropriate military initiatives, colonies have not, in fact, taken permanent root.

Once colonies had been established and proved (or appeared) profitable, effective military presence of the mother country was also essential to discourage competing colonial powers from taking over in the name of true religion, civilization, manifest destiny, and other covers for imperial design. Whenever military supremacy of the colonial masters was allowed to fritter away or disappear abruptly (e.g., Spain in the early 1800's, or France in the last decades of monarchy and immediately following the Revolution), the colonies either acquired independence on their own or were given new masters by invaders.

We have here at some length elaborated on the genesis and the causes for stability in the colonial regimes which in the past have ensured continuing growth and thus reasonable political stability of the Western world. To be sure, there were in the West many remnants of religious and dynastic conflicts; class struggle came to the fore in the mid-1800's but nonetheless the basic mechanism for securing markets, increasing production, and accumulating wealth was present and has appeared almost as stable as the laws of nature ... mostly to the beneficiaries. It is now necessary to examine whether this state of affairs had any chance of enduring and if not, whether the causes of disappearance are fundamental and permanent.

* * *

The "era of trade beads," so rewarding to daring entrepre-
neurship, is definitely over. The managers of the less-
developed countries (LDC's),[7] quite irrespective of race,
creed, color, or national origin, are perfectly capable of
watching stock and commodity exchange tickers; in point of
fact, by and large they do so. Thanks to the blessings of jet
travel, they visit Western metropolises quite frequently;
their sons (and quite significantly, their *daughters*) are often
among the most favored, if not the most assiduous, segments of the Harvard, MIT, Yale, or Stanford student
bodies. Even illiterate classes in the LDC's are superficially
aware of Western values and life styles thanks to the
friendly information services offered by the USIS[8] and to the
not-so-friendly propaganda beamed at them in profusion by
the Soviet camp. Assuming that some new substance, highly
prized in the West, would be discovered in some remote part
of the world, and assuming further that it could be
recovered at nominal cost, the situation (in sharp contrast
to the early colonial period) would no longer offer any real
prospect for profitable and long-enduring trade relationships. The producers would soon discover the "replacement
value" of their exports; they would combine into cartels or
monopolies whenever possible, or at least would cleverly
bargain and play one purchasing nation against another.
They would exact whatever the highest price the traffic
would bear and rush to invest their new wealth into their
own modernization, but also (preferably) into profitable real
estate and business properties under the protection of the
Western nations. Owing to their fast accumulating wealth,
their religion and philosophy would attract world-wide
curiosity, if not automatic endorsement. Even the super-

---

[7] Word images, as always, carry powerful emotional messages. No longer is it
fashionable to speak of primitive, backward, or poor nations; even the
term underdeveloped has unpleasant, ontological connotations. The situation is precisely congeneric to the evolution of terminology among the
U.S. social workers: the poor have been transmuted by steps into the
destitute, the deprived, and eventually, the underprivileged. The next step
is obviously the "LPP's" (less privileged persons).

[8] See Glossary

powers would give respectful hearing to the political views of their rulers, at least until some local upheaval conclusively established (*a posteriori*) their lack of wisdom.

Much as we may regret it, religious beliefs have ceased to be valuable Western export commodities. There might be some room to speculate as to whether or not the Christian faith would have ever been successful in penetrating other cultures (especially those with deeply rooted civilizations of their own), had it not been sustained by the example of power and by the real presence of military threat. [9] To depend on the intellectual and spiritual elements in our culture to form the basis of leverage in future dealings with Third World trading partners is probably quite unrealistic. In further support of this point, one may observe that, even though our public morals are ostensibly derived from religious ethics, the Western governments are professedly secular and insist on rather strict separation of church and state. Neither can our foreign trade policies be accurately described as being in substantial conformance with Christian principles. As to exporting the ideas of democratic government, wherever it has been genuinely attempted in the past (mostly by the post-World War II U.S. missionary spirit), the success has been far from spectacular, widespread, or enduring. [10] The wisdom and statesmanship demonstrated by Western leaders in recent decades, the stability and promise of our political institutions, do not constitute examples that would invite unquestioning acceptance by potential Third World followers.

*

[9] The spiritual impact of the many devoted missionaries should not be underestimated. On an individual basis, they often transmit (and have done so in the past) the ideological quintessence of the Western civilization without mercantile interest or objective.

[10] Japan is perhaps the single major exception following World War II, but then again, it is a unique situation. Japan has never been a colony; it has traditions going back to more than a millenium of deliberately and astutely borrowing and adapting from other cultures.

Considered in its totality, the military power of Western nations has also ceased to be an effective instrument for the protection, let alone the coercion, of former colonies. More generally, Third World nations can no longer be effectively coerced into economic subjection by the use or the threat of military force. Traumatized by two world wars, Europe is no longer morally equipped to sustain overseas military ventures, especially when the outcome may be dubious and the net economic rewards questionable.[11] Minor paratroop-type operations, serving to police the less stable mini-countries of Africa, might still be undertaken mostly for reasons of nostalgia, prestige, or influence-seeking, but certainly not in view of economic domination. The U.S., still fascinated by the specter of communism, smarting from the Vietnamese experience, and influenced by its domestic racial dilemmas and anticolonialist traditions, is not in position to exert direct military pressure in support of essentially mercantile undertakings. In fact, well before the Vietnam war, the United States had, as a matter of principle, formally renounced the use of force in order to further its economic interest and had actively discouraged others, friends or foes, from doing so.[12] Significantly, all Western nations in recent years have changed the name of their War Departments; they are now operating under names such as Ministry of Defense, National Defense, etc.; the Japanese go so far as to insist on Ministry of National Self-Defense.[13] None of the nations of Western Europe, Japan or North America publicly suggested intervention in Iran in 1978-79

---

[11] In general, an overly cynical observer might suggest that the moral aspects of military initiatives are only questioned when they end up in defeat or when the tangible rewards in terms of prestige, wealth or influence are clearly marginal.

[12] Declarations by John Foster Dulles in the wake of the Suez crisis of 1956 when the U.S., on President Eisenhower's explicit orders, turned against our own European Allies.

[13] Will, within the next few years, Orwell's "Ministry of Peace" become a commonplace designation?

when their vital oil supplies were clearly threatened by the demise of the Pahlavi dynasty.[14]

The question of whether or not continuing capital investment for resource and infrastructure development projects can remain a major tool in Western hands requires careful examination.   Until about 1970, following the peak of decolonization, the retreat of Great Britain from East of Suez, and the height of the U.S. Vietnam involvement, direct use of military power or indirect influence exercised through reliable allies have remained effective means to protect traders and investors.[15] Even in the absence of overt military domination, investment was protected by a combination of political influence and the threat of economic ostracism in case of behavior contrary to that accepted by Western business practices.   In the post-war period Japan and Europe, without significant military strength of their own, have confidently acted under the umbrella provided by Western-dominated economic structure.   Even at the present time the petropowers of the Middle East depend on the Western economic order.   Should it weaken or disappear, they will have to restrict their investments into local development or that of ideologically kindred nations, but certainly not in areas where their assets might be threatened by political instability.

In the absence of military might or resolve, the dominance of the global economic system may still ensure reasonable

14 Whether or not any Western nation or coalition would actually intervene with military force to protect its oil supplies is not clear. The brave talk of unilateral intervention by the U.S., openly discussed in 1973, has by now largely subsided. Except in the case of a clear and blatant aggression against vital U.S. interests (which our potential adversaries would be most careful to avoid), we would probably hesitate and temporize. (Note added in proof: These words, written in August 1979, amazingly still hold true in May 1980, after the tumultuous events in Iran and the Soviet thrust in Afghanistan.)

15 The classical examples of the more distant past are the East India Company and the Hudson Bay Company, both chartered for joint military and commercial enterprises. In more recent times, the powerful U.S. influence in Latin America has encouraged Western investment there.

safety for capital investment in the Third World but, when attempting to assess this mechanism in terms of its long-term leverage to secure critical raw materials under advantageous and dependable conditions, two further factors must be taken into consideration. First, the intrinsic availability of Western capital for investment (with the possible exception of Japan) is open to question. The rate of capital formation is painfully slow, barely adequate to satisfy the need to continually modernize their own produc-tive machinery, to develop their own resources, or to maintain their infrastructure in keeping with the needs and the growing desires of their population. Second, both the availability and safety of Western investment capital might be threatened by the systematic and relentless economic pressures exerted by competing societies. While threats by the well-publicized Soviet military buildup impose no inconsiderable burden of armaments, the quasi-endemic political turmoil abetted, when not deliberately fomented, by the Soviets or their proxies constitutes a further element of risk in long-term capital commitments.

On balance then, while the investment mechanism has been powerful in the past and still carries considerable momentum, its future, based on extrapolation of clearly observable trends, is definitely bleak. The further rise of "multinational" companies, which are an operational means for implementation, rather than a basic mechanism, would not change this picture.

Several Western countries vitally dependent on Third World raw materials have subliminally or explicitly sensed the gradual erosion of their trade leverage. As a form of rear-guard action, many have borrowed a time-honored prescrip-tion from the ancient art of exercising political influence through the supply of arms. When the respective political goals of suppliers and recipients are reasonably stable and have essential common elements, weapon trade is probably justified. So it was quite natural to enter the arms trade as an element of economic as contrasted to political leverage. Frequently, profit opportunities (or expectations) for Western private enterprise are huge and serve to conven-iently support the rationale underlying favorable arms trade policy decisions.   The disappointing fact, nonetheless, remains that the arms trade is simply not an adequate means to gain long-term influence (political or economic) unless the recipient's political goals are stable, deeply engrained, and kindred to those of the suppliers. For one thing, modern weaponry is quite expensive, especially if it is to compete effectively in the recipient region. Owing to local rivalries and reasons of prestige, once started, the momentum of weapons buildup is most difficult to reverse, especially when the recipients are in position to play one supplier versus the other, capitalizing either on the East/West antagonisms or simply on the competition among Western suppliers. Eventually there is obvious danger of polarization and radicalism of highly militarized regions. Our relative capability for effective military control becomes even more costly and less promising. Weapons may be interesting *adjuncts* to the basic driving forces of trade, but they cannot be a lasting substitute. They may be useful to start the initial process, but in the long term the Western nations, if they elect to use the supply of armaments as the basis for countinuing interaction, are almost certain to lose whatever leverage they originally may have possessed.

We have argued up to this point that a multilateral, balanced trade flow following the economic and cultural gradients is essential to the preservation of high standard of living and stability of the Western nations.   We have observed that all the traditional leverage factors: cultural, economic, technical, and military, have disappeared, have been voluntarily relinquished, or have been allowed to erode. The Western nations must find ways to reinforce the remnants from the traditional trade mechanisms and eventually define, develop, and exploit *new mechanisms* if they can be found in time and offer reasonable promise of long-term validity.

The solution to this problem is far from simple; we do not live in a rational world, neither can we say that the Western nations are the only competitors.   The Western societies face, in addition to the erosion of the economic basis of their survival, major ideological challenges supported by rather credible military developments. These are the ones we must turn our attention to now.

*** * ***

Indeed, the Western nations[16] are no longer the only group which claims or believes that it is (and is to remain) the foreordained leader of the world.   Other, vigorously competing groups have arisen to challenge this claim and the implied conclusion that the West is entitled, on that basis, to a disproportionately large global share of resources, wealth, and rewards.

Challenges of this type are part of the normal evolution of any society; as long as the dominant institutions are viable and adaptive, they may even benefit from the ongoing questioning of their legitimacy.   In the first half of the twentieth century, anarchism, socialism, communism, fascism, nazism, falangism, and some others have come to the fore and have attempted to redefine the relationship between the modern state and the individual.   Special emphasis, in general, is given to the concentration of executive and legislative power, to the central control of the economic activity, and to the explicit subordination of the citizen's individual rights to the interests of the state. Owing to the special combination of leadership and social conditions following the First World War in Russia and the Second World War in China, the communist ideology has succeeded in creating powerful governments, showing remarkable endurance in spite of the violent convulsions of the first few decades. The Soviet regime has by now been in power for more than two generations; it has survived (with considerable material help from the West) the Nazi onslaught of World War II and has achieved superpower status recognized by all nations, including its main rival, the United States. The regime of the Peoples Republic of China (PRC) has been in power for more than 30 years, following several decades of anarchy, foreign invasion, and civil war. In spite of considerable initial difficulties, it has apparently succeeded in offering acceptable living standards to its population, now approaching one billion in size.   Its long-

---

[16] In this context they include Western Europe, The U.S., Canada, Japan, South Africa, Australia, and New Zealand.

term military potential is recognized as a serious threat by the Soviet Union; its alliance and friendship are eagerly sought by the U.S. under the last three Administrations and by virtually all other nations.

The growing challenge to the West by the socialist/communist ideology is one of the central facts of the second half of the twentieth century; it is powerful, multipronged, and fundamental. It is essentially decoupled from tactical moves such as shifts in alliances, detente, cold war episodes, and opportunities for hostile probings at the periphery of the respective spheres of influence. Added to the objective factors which we just have discussed, it forms the ideological basis for the steady erosion and the eventual elimination of the world-wide Western economic domination.

There can be no doubt about the reality and the momentousness of this challenge. The political doctrines of both the USSR and the PRC explicitly assert that the class struggle, which has resulted in the victory of the working class in their own societies, is now being repeated at the larger scale on the global scene. The misery of the large masses of toilers around the world is, according to this doctrine, a direct result of past and current criminal exploitation by the wealthy nations. Such exploitation, still according to the doctrine, is unjust and must be eliminated by "peaceful" means if possible, but by violence whenever necessary or opportune.

As long as such assertions were propounded mostly inside of communist boundaries, as long as we could complacently observe the many difficulties and distortions of the socialist economies, or point with self-righteous horror at their human rights violations, their imperialistic and cynical efforts to subjugate smaller nations at their own periphery, the West could assume that the very nature of their own superior economic system will, in the long run, attract the enthusiastic support of the Third World nations. It may also, according to the Western belief, encourage them to work peacefully toward a common future (i.e., without disturbing the comfortable premises wherefrom the Western economic system continues to derive its increasing wealth). The stark reality is, however, quite different. Thanks to the much-improved technical means of communications[17] the messages deliberately tending to promote global class struggle reach now the humblest village in the most remote corner of the earth, in the very own language of the villagers. The communist economies may appear backward or inefficient by Western standards, but they look miraculously productive in the eyes of LDC observers, especially of those who virtuously reject consumer luxuries in order to advocate a more equitable distribution of vital necessities. The many internal human rights violations are easily condoned by people whose history has not offered them long periods of civil rights guaranteed by Magna Carta, Habeas Corpus, and other typically Western-cherished devices; and their leaders are only too prone to apply dictatorial methods to increase productivity, to enforce conformity, or simply to conceal their mistakes.

The psychological aspects of related arguments are also very much in favor of the socialist viewpoint. It is relatively easy to argue that abject poverty and inhumane sufferings are the result of past injustice and cruelty; that the victims are entitled [18] to speedy reparation, and that there is an easy way to acquire riches, namely, to take it away from the wealthy. By way of contrast, it would be quite difficult to sustain (truth notwithstanding) that the domination by the West is basically due to the sharp cultural differences; to the unique war-like virtues of its people at the past time of rapid ascendancy; and to the fortuitous and felicitous combination of industrial development and political institutions rewarding individual initiative. Whether genetic factors have also played a role in the past has not been proven, and, at any rate, may not be necessary or politic to argue.

---

[17] See Chapter IV.

[18] Note that even the terminologies are uncannily similar to those used in domestic social debates.

In the struggle for retaining its supremacy, the West should derive little comfort from the much discussed Sino-Soviet split.   The conflict between the two communist blocs is serious and may well culminate in a major war, but they both take fundamental issue with the legitimacy of the current West-centered global economic order and reject any justification for its continuing future.   The relative economic weakness of China, its present tactical acquiescence of the need for modernization, with the concomitant need for playing the "U.S. card," or for its possible desire to participate in some triangular grand strategy game, alternately oriented against the U.S. and the Soviets, are all important factors for the near term, but should not lull the West into a false sense of security.   They simply indicate that the Soviet Union is, for the time being, the senior participant in the ideological and economic struggle against Western domination.

One of the characteristic components of the hostile ideological efforts promoted by the USSR is aimed at creating confusion in the minds of the West. The objective is to influence the Western intelligentsia by promoting self-doubt and guilt feelings, by encouraging the no-growth mentality, and by playing on our Christian compassion in regard to the stark contrasts of the starving multitudes and their untold misery in the Third World against the conspicuous consumption in the West. These, added to many other similar messages, are aimed at intensifying and propagating the stresses already present within the Western nations.

The Soviets and the Chinese have both been remarkably astute in initiating confusing, competitive, or potentially threatening economic actions. Thus, the Soviets conclude, whenever possible, long-range commercial treaties with key nations such as West Germany and Japan, creating thereby ambiguity in their political allegiances. By shrewdly selecting the manner and timing of their huge 1972 grain purchases in the U.S., the Soviets even managed to satisfy their own urgent need to supplement the cumulative deficit in their agricultural production, while causing at the same time significant acceleration of inflation within the U.S. as well as world-wide.[19]

The Soviets are in position to control the supply of some essential materials such as tungsten and chromium.[21] They are the world's second largest producers of gold and since their domestic market is essentially state controlled, they can use their gold hoard to finance their foreign trade when necessary, or to disrupt the global gold market whenever they choose to do so.[20]

As an example of the long-term, multi-pronged aspects of the Soviet economic   initiatives, one of their most fascinating and ominous investments is the creation in the past 20 years of a modern merchant fleet covering most of the world's major shipping routes and practically all the attractive internationally available fishing areas.   Such a state-directed,   major   long-term   capital   investment program would be unthinkable in peacetime in the U.S. where the flow of capital is mostly controlled by private profit expectation.[21] The Soviet's entry as a major participant in the global shipping and fishing industries has already paid handsome dividends and is quite likely to do so in the future as well.   It offers all the conventional economical benefits accruing from overseas trade, access to raw materials, markets, and revenue from shipping.   Its modern

---

[19]   It is interesting to observe that in 1972 and then again in 1973 at the Bucharest Conference on world-wide food shortage, the Soviet market actions were left essentially undiscussed. In contrast, the U.S. received severe blame for food shortage and high prices.

[20]   If further incentive be needed for the Soviets to aggravate the Western inflation rates, the appreciation in their accumulated gold reserve would offer ample justification. Since 1971 the dollar equivalent of the Soviet gold has risen at least by a factor of 5, but possibly by a factor of 8.

[21]   Japan, and to a somewhat lesser extent Western Europe, have state-controlled or at least state-encouraged, mechanisms to steer investment capital into the desirable industrial sectors. In the U.S., in spite of the huge and manifold federal subsidies to the maritime interests, national security considerations are mostly a pretense, at best an afterthought.[22]

fishing vessels and sea-going processing factories provide important nutritional supplement to the Soviet population. The abundance of seafood can offer leverage in the protein-starved LDC's; experiments with industrial processing of commercially undesirable fish species appear promising.[22] The availability of ample shipping tonnage allows the Soviets to offer credible presence and support to their overseas friends and allies; in several recent instances sea lift and maritime lines of communication of most respectable capacity even by Western standards were provided to Soviet subsidized proxies in Africa. By manipulating at will the shipping rates, the Soviets are in position to disrupt the Western shipping industry with the axiomatic consequences (in nations where investment is still at the discretion of individuals and private corporations) of discouraging shipbuilding and the long-term deterioration of shipyards and shipbuilding industry. Thus, both the capital equipment and the capital facilities, required to upgrade the equipment in times of urgency, are allowed to decay; the available pool of skilled manpower needed to replace and to operate the ships is being dispersed into more rewarding sectors of the economy. These trends are almost impossible to reverse in less than a decade. By way of contrast, the Soviets and their allies benefit from a modern, growing seagoing fleet; they have excellent shore-based construction and support facilities, they offer substantial encouragement to the recruitment and formation of skilled construction and operating personnel, all automatically available at any and all times for the purposes of the government. Even in peacetime, Soviet vessels systematically participate in oceanographic data collection and in surveillance and intelligence activities in essential support of their military establishment. At the global strategy level, the Soviet maritime investment challenges the very structural basis of Western supremacy—the ownership of effective economic and military links between continents.

---

22   The Soviets are one of the major nations still engaged in large-scale whaling operations. The criticism by the protectors of wildlife so vocal in the West is remarkably muted on the subject of Soviet activities.

A number of politically inspired "economic" initiatives are directly aimed at the Third World nations. They, in sharp contrast to the corresponding Western practices such as pursued by the World Bank and the Agency for International Development, play down monetary interests and concentrate on a few spectacular projects highly visible to the population. Typical among these are participation in major development projects, such as the construction of the Aswan Dam by the Soviets, the construction of steel mills in India, and the capital support and labor help by Red China in the construction of the Tanzanian railroad. The Soviets, whenever necessary, do not hesitate to underwrite uneconomic ventures; the Cuban example proves that the relatively small yearly subsidy (by U.S. standards) yields disproportionately large rewards in terms of foreign policy leverage, especially in Latin America where the potential exploitation of the anti-Yankee feelings appears to offer remarkably fertile ground for their anti-imperialist campaign.[23] More generally, the Soviets and the Chinese have managed to posture themselves as the champions of national liberation movements, thereby accelerating the decolonization and creating turmoil within and among the newly emerging nations. The Soviets have shown considerable far-sightedness in their efforts aimed at "educating" the revolutionary elites among the youth of LDC's; "Lumumba University" in Moscow, established in the mid-60's, is in remarkable contrast to our corresponding efforts, which mostly cater to the sons and daughters of the wealthy and privileged strata. We do not hesitate to include, often to our chagrin, training for military officers associated with notoriously repressive regimes.

More generally, we must face the fact that, helped by the pressures of our tradition and our fascination with private interests, we seem to follow rather consistently a somewhat short-sighted policy in that regard. The U.S. in particular manages to be almost always on the side of conservative,

---

[23] (Note added in proof) It has taken nothing less than a change in the U.S. Administration (January 1981) to recognize publicly the import of the Cuban foothold in Nicaragua, Guatemala, and El Salvador.

dictatorial parties. Only very recently have we taken an unambiguous stand in support of the black majorities of Africa. It is no wonder that, as a matter of routine, the majority of the United Nations is now opposed in most instances to the West and hardly ever fails to exploit opportunities to denounce Western economic domination.

None of the foregoing should be construed as suggesting that the Soviet-inspired ideological and economic drive is precisely orchestrated according to some well-thought-out master plan. Its broad strategic objectives have been decided, but specific moves for implementation are quite flexible and opportunistic. It is greatly helped by the state of almost permanent confusion and conflict in the West, as well as by an almost complete lack of long-range policies or constructive initiatives. In all fairness, though, "the wicked are not always clever"[23] and quite often Soviet moves backfire or seem to work against their own long-term interests. Egypt (1972) and Somalia (1978) are among the recent adverse outcomes of short-term Soviet commitments. Their support of the idea of "global redistribution of wealth" may prove hurtful to their longer term interests when their own industrialized society reaches the limit of internally available basic resources, which is predicted (according to various sources) to occur in the mid to late 1980's. Much along the same lines, the vigorous championship by the Soviets of wars of national liberation may be taken somewhat too literally by their own fast-growing racial or religious minority groups.

*

Without a basic understanding or, better, without a united front among the Western nations aimed at collectively preserving their overall preeminence in the world economy, even short of dominance; without a plausible action plan to implement imaginative long-term policies to this effect; there is no reason for hoping that the rapidly accelerating decline in the relative position of the West will be reversed, stopped, or even slowed down. In past times, military force might have been employed to redress the situation, but we have well noted that many impediments eliminate that approach for most practical purposes. The nature of modern warfare, to be discussed in the next chapter, further suggests that employment of military force to support economic objectives will become an increasingly desperate and hopeless proposition in the future.

CHAPTER VIII

Threat and Use of Force

THE FACTORS WHICH HERETOFORE FAVORED WESTERN STRATEGIC SUPREMACY HAVE CEASED TO OPERATE. NO LONGER CAN THE WEST USE MILITARY FORCE TO PROTECT OR TO REGAIN ITS IDEOLOGICAL AND ECONOMIC DOMINATION—BUT ITS ADVERSARIES CAN AND DO ENVISION ITS USE FOR THEIR OWN PURPOSES.

Moral considerations are brought to bear against the use of force mostly by people who are in position to protect their interests by means which are, in their opinion, far more conducive to that end. The threat or the actual use of force has been since time immemorial the coercive basis or backdrop of all human conflicts; we may quite confidently surmise that this will not perceptibly change in the foreseeable future, regardless of the peace-loving rhetoric around the world.

The three major wars which largely defined the political complexion of the West by the late 19th century[1] have all resulted in perceptible and lasting benefits for the victors. Even the vivid news coverage (by standards of the time) of overseas conflicts such as the Crimean War (1854-1856), the Boer War (1896-1902), the war in Sudan (1896), the Korean War (1896), and the Russo-Japanese War (1904-1905) could not erase from public consciousness the belief that wars are,

---

1 The Holy Alliance vs. Napoleon (1803-1815); the U.S. Civil War (1861-1865), and the Franco-Prussian War (1870-1871).

on the whole, immoral but necessary, and that they are beneficial if properly conducted and victorious. They also offer gainful occupation to the talented, adventurous youngsters, mostly of good families. Kipling, Lyautey, and Churchill typified the spirit of the times at the turn of the century.

The World Wars (1914-1945 with a brief interlude between 1918 and 1939) have thoroughly shattered this view in the public mind as well as among political and military analysts. Taking the long view over half a century, Germany and Japan (the major losers) are today quite better off than their "victorious" adversaries, England and France. Following the Second World War, the Soviet Union and the U.S. came out on the winning side, but the true gains were shortlived. In order to satisfy its historical need for buffer states, the Soviet Union was driven to close association with a sullenly hostile satellite empire, which may bear the seeds of future disintegration of the Soviet political concepts. In the East, following Stalin's death, the Soviets permitted (or may have provoked) the rise of the Peoples Republic of China, a major ideological competitor, which in the longer run may become a serious military competitor as well. The U.S. ended the war as the world's sole nuclear "superpower," but in the process it permitted the rise of the Soviet Union which, in less than a decade, was in position to challenge the U.S. monopoly. By the mid-1960's the military supremacy of the U.S. came to be questioned, at least in some regional contexts, and by the late 1970's the U.S. was earnestly negotiating to retain "parity," "equivalence," or at least "sufficiency."

In the immediate, short-term perspective, World War II accomplished its stated purpose. It eradicated the Nazi dictatorship aimed at the hegemony over the Eurasian continent and destroyed the basis for Japanese military domination of their rather ambitiously defined "co-prosperity sphere." But in the longer term, far from making the world safe for democracy, the result of the two world wars of the twentieth century may well be summed up as a major strategic setback for the "First World," i.e., the major

Western-style industrialized nations [2] and, to a somewhat lesser extent, the "Second World," i.e., the Soviet Union and some of its satellites. In economic terms, these wars have consumed a large fraction of the accumulated capital assets of the West; they also have given further impetus to the disintegration of the colonial system and thereby contributed to the destabilization of the world economy. In military terms, World War II demonstrated the practical use of nuclear weapons, starting thereby an arms race of which we can not yet see the end; and creating a world-wide military-oriented research and development establishment, as an institution with its own momentum and with the associated military and economic consequences. Perhaps even more importantly, in the aftermath of the two world wars and of their immediate sequels, the Western societies, at variance with their adversaries, have developed an increasing, visceral reluctance to consider military action as a desirable, or even acceptable, means for conducting foreign policy. These military and sociopolitical consequences are examined in this chapter, with some emphasis on their asymmetric impact on the two major centers of influence, conveniently (even though inaccurately) described as the "West" and the "East."

* * *

The brief history of the nuclear age is a tragic illustration of penalties paid in the longer term for ill-advised, or rather incompletely explored, decisions offering in appearance considerable near-term pay-off. There was almost unanimous approval of the use of atomic bombs in 1945 over Hiroshima and Nagasaki. President Truman was warmly praised for having saved American lives by dramatically shortening the war. A short while later, we attempted, through pleading and negotiations, to control the spread of atomic weapons, but by 1947 it became obvious that the

2 Once more, Churchill's words, written in 1953, sound prophetic and ominously foreboding... "How the Great Democracies Triumphed and So Were Able to Resume the Follies Which Had So Nearly Cost Them Their Life"... (theme of Volume VI of his Memoirs).[23]

Soviet Union was actively pursuing related research programs. Aided or not by espionage in the West, the Soviets exploded their first atomic device in 1949; by 1954 their progress in thermonuclear weapons justified the U.S. commitment to development of ever more powerful weapons of this type, which at that time were a prerequisite for efficient delivery by intercontinental missiles.

Up to that point, or perhaps until the early 1960's, our course of action could have been questioned on basis of moral considerations, but at least it contained a core of military logic--so long as we retained unquestionable supremacy in nuclear ("strategic") weapons. Many knowledgeable people question whether or not such supremacy could have been preserved in the long term, but sometime during the 1960's a decision was taken, or at least implemented which, deliberately or inadvertently, resulted in its gradual erosion into superiority, then, by degrees, into parity, essential equivalence, sufficiency, and into other "codeword"-described postures, all connoting decreasing U.S. nuclear strength in relation to the Soviet Union.[3] To have introduced nuclear weapons into the panoply of modern warfare and, after having been unsuccessful in the attempts at controlling their spread to other nations, to have viewed with equanimity our dedicated enemy, the Soviet Union, to acquire an equivalent nuclear power may well be described as a mistake of truly historical proportions. At the present time, the nuclear "balance" is at best precarious; many Western observers find that it is in fact shifting in favor of the Soviets, arms limitation treaties notwithstanding, in view of the relentless momentum of their strategic weapons investment. The erosion of our nuclear power has not remained unnoticed by our Allies. The relative lack of cohesion in our major alliances, such as NATO and the

---

[3] The identity and motivation of those responsible for this "decision" is not essential to our discussion, but it was never publicly discussed or approved by Congress. To the contrary our leaders (in public) enthusiastically supported the solemn commitment that "our defenses are, and are to remain, second to none..." There are reasons to believe that a few more "liberally" oriented among our policy-makers in the early '60's have deliberately promoted nuclear parity as an element of peace, as contrasted to letting peace rest on the unquestionable and unquestioned U.S. supremacy.

Japanese defense treaty; the demise or quasi-demise of CENTO and SEATO; the lack of effectiveness of the OAS, amply prove that we have no longer any real leverage on our purported friends and Allies. Perhaps they see the evolution in the balance of nuclear power with less biased eyes than those of the U.S. Government. Neither is the meaning of the shift in nuclear balance lost on the rest of the world; the potential for influence, pressure, and eventually blackmail could appear irresistible to the Soviet leaders. A bitter, ironical ring is associated with the pronouncement of the late Sir Winston Churchill to the effect that "our future security may well be based on the balance of terror." He would never have envisioned the possibility that we could jeopardize our security by failure to maintain a true balance, preferably with a comfortable margin for errors systematically in our favor.

The vastly changed technical aspect of warfare, caused by the advent of nuclear weapons, is one of the major reasons why large-scale wars are perceived as unattractive in the West by the public and the policy-makers alike. Other reasons may be traced to the inherent vulnerability of industrialized societies, which has entered the public consciousness as much through the threatening propaganda of their adversaries as through the equally well-publicized views of their own experts. Western societies are by far more vulnerable physically at the present than at any other time in history. The central and obvious fact is that long-range missiles, launched from open oceans, from space, or from far-away continents can reach any point of the homeland. Coupled with high-accuracy, space-based reconnaissance, any military target, including hardened underground facilities, can be destroyed by means of long-range nuclear-tipped missiles in a matter of minutes with high degree of certainty. Cities, industrial and "infrastructural" targets can be obliterated with equal facility at any time of the attacker's choice. The exact destructive mechanisms and those lethal to the population are matters of interest to the experts, but there is no argument among them about the belief that a deliberate nuclear attack on the major urban centers would lead to several ten million people being killed and most of the target areas destroyed and uninhabitable. Even an attack conducted with nuclear weapons of "surgical" precision exclusively aimed at strategic targets, such as missile launch sites, airfields and submarine bases, would leave hundreds of thousands or perhaps several millions of people killed with massive damage to adjacent civilian areas.

Against such threats a modern industrial society, such as the U.S., has no meaningful resilience whatsoever. The high degree of industrialization, the extreme dependence of urban areas on transportation, utility, communication interties with, and among, distant regions makes the U.S., as a type of industrial society, particularly vulnerable especially when no serious plans have been implemented in order to ensure protection of the population, of the industry and, more generally, of the organization, equipment, and infrastructure which may permit eventual recovery.[24] Equally lacking is any psychological preparation of the population to face emergencies of such order; the tendency of exaggerating the damage, of exploiting its sensation value and of reacting in an uncontrolled panic mode may all act in concert to amplify the intrinsic vulnerabilities. Peacetime attempts to impose discipline, preparation, or to suggest correspondingly appropriate attitudes are resented and resisted in the name of freedom; advocates of civil and industrial preparedness even face the risk of being accused of warmongering or worse.[4]

---

[4] The reaction of large urban populations to catastrophies at such massive scale can not be predicted on basis of past history. The last natural disaster (puny by the standards of projected nuclear warfare) affecting a modern city was the San Francisco earthquake in 1906; the evidence can be read in both ways; large-scale panic and widespread heroism were inextricably intermingled. The bombings of Hamburg, Dresden and at a much smaller albeit destructive scale, of Rotterdam in World War II, have apparently not caused massive breakdown in the social order. The behavior of London during the 1940-41 Blitz has been often and admiringly described as exemplary. On the other hand, the current cinematographic fascination with catastropic themes and the massive emotional reactions to relatively benign events, such as nuclear power plant malfunctions, offer support to the opposite view, i.e., that the problem of maintaining order and discipline might become one of the major contingencies.

Interwoven with the lethality of nuclear weapons and the vulnerability of modern societies are the relatively new aspects of large-scale warfare related to the *tempo* of operations. Future wars, especially those which involve the use of nuclear weapons against the superpowers' respective homelands, will possibly reach their peak combat intensity within a matter of days (possibly hours or even minutes) following the inception of hostilities. "Strategic" warning, based on intelligence, may offer premonitory signals a few days ahead of time, but this is by far insufficient for implementing any but the most urgent military posture changes. It is perfectly conceivable that the outcome of the war would in fact be determined in a few days or a few weeks, but at any rate within a period much shorter than that which would be required for meaningful industrial mobilization. Thus, not only may the military posture be found wanting at the outbreak of the war, but the major traditional strategic advantage of the industrial nations, the ability to convert their superior industrial capacity into effective support of the warfighting effort, would be essentially denied to them in this type of conflict.[5] Even if the war does not reach its peak intensity shortly after inception, the performance and the survival of the war industries and of the civilian economy in a protracted nuclear combat environment are problematic, to say the least.[25]

Few people have attempted to think through (let alone analyze in convincingly competent detail) the initial phases,

[5]   The Western nations (and more particularly the U.S.) can no longer depend on mobilization of their large civilian/industrial base to support protracted wars with a view of winning decisive advantage. In previous conflicts, precisely because of its size and remoteness, the continental U.S. territory was essentially not accessible to the then available weapons. The Allies, in spite of having been subjected to fearsome dangers by the initial onslaught of well-prepared military machines of the dictators, had more than a few years to bring their overwhelming strategic superiority to bear on the outcome. In World War I, the United States did not enter the conflict until 1917 (almost three years after its inception), and except for minor shipping losses, was immune to enemy attack. Similarly in World War II, the U.S. mainland was still essentially invulnerable to enemy attack and had two to three years on hand to prepare its army and to convert its industry into the "arsenal of democracies," which ultimately carried the decision.

the conduct over protracted periods, the termination and the recovery following major wars involving the superpowers. Even fewer are those in position to delineate the problems of "competitive societal recovery," i.e., the reconstruction of social, political, economic and military components of the survivor nations in competition (or even in a state of overt, endemic hostility) with their erstwhile enemies and allies.[26]   Considerations of this type, far broader and more general than those reflected in the currently available "nuclear exchange models," should be used when attempting to assess the relative strengths and weaknesses of a nation in the perspective of nuclear wars. But even in absence of adequate knowledge, the foreign policies of all Western nations have accepted since 1945 the basic premise that nuclear wars should be avoided at almost any cost. So engrained has become this attitude until very recently that the idea of fighting major nuclear wars remained literally unthinkable, at least among the Western civilian policy-makers. Whether or not their Soviet counter-parts genuinely share this conviction is perhaps the most momentous problem which faces our generation.

* * *

The invention of nuclear weapons is the major, but certainly not the only, event which brought radical changes to the concepts of warfare since World War II. Public view is bedazzled by the spectacular, well-publicized and emotion-ally novel scale of nuclear warfare; megatons, fireballs, and invisible lethal radiations have a way of captivating people's imagination which has no equivalent in recent human experience. While obviously critical to the outcome of any particular military engagement, the physical aspects of nuclear warfare should not hide from our view other associated consequences perhaps even more important from the social viewpoint.

The sudden and decisive use of atomic bombs against Japan in 1945 has deeply seared into the public consciousness the vital importance of being technically prepared, protected against adverse surprises; of being there "first." Spine-chilling scenarios have been published as to what could have happened had the Nazis or the Japanese won the "race" to possess the atomic bomb. The frightening scale of atomic destruction has driven home, as nothing else could have done, the belief that catastrophic surprises are possible and that lack of military preparedness might be tantamount to a nation's death warrant.[6] This may well be one of the important driving forces behind the arms race which by now has continued unabated for more than thirty years among the superpowers and their respective Allies. Much for the same reasons, and in spite of all the laboriously negotiated restrictive treaties and test bans, proliferation of the ownership of nuclear weapons appears to remain a perma-nent trend in international life. Current membership in the "nuclear club" stands at six or seven;[7] with others impatiently knocking at the door. The proliferation process is slow and some of the most important components are surreptitious, but we should not blind ourselves to its reality and future consequences. Well before the end of the century additional nations, estimated between 10 and 20 in number, will be in possession of weapon-compatible nuclear devices together with competent delivery mechanisms. Whether or not under those circumstances the stability of international relationships will be easier to preserve than in a non-nuclear world is a question which is at least debatable.

An important by-product of the arms race is the institution-alization of military research and development with the concomitant ritualistic (upward) revision each year of the hostile threat which any one of the advanced nations might face. This is not the place to argue the morality or the effectiveness of continuing large-scale military expendi-

---

[6] Curiously, the idea of strategic preparedness has clearly entered the national consciousness (at least in the U.S.) mostly in relationship to offensive power. Defense, and more particularly civil defense, has not received nearly the same degree of public acceptance or official sanction in the United States.

[7] U.S., U.K., USSR, France, PRC, India, and possibly South Africa, and/or Israel.

tures, but at least one essential consequence must be noted: by stimulating the growth, both in size and technical sophistication, of a military-oriented technocracy the arms race has spawned major advances even in weapon systems only remotely associated with nuclear weapons. The cumulative impact in the last twenty years has been truly revolutionary. The advent of sophisticated "conventional"[8] missiles has fundamentally changed the nature of land, air, and naval warfare, requiring corresponding changes in combat equipment, aircraft, and ships, as well as in ancillaries such as reconnaissance or surveillance vehicles, including space satellites.

While conceptually and potentially more powerful, the defense establishments of most Western nations also suffer from almost chronic built-in dysfunctions: the rapid proliferation of weapons of different types contributing to the same or similar military missions and the frequent and repeated upgrades to incorporate the latest technical advances. All these lead to standardization and inter-operability problems, they all contribute to a steep rate of increase in operation and maintenance costs; they also lead to increasingly stringent personnel training requirements. The institutionalization of military research and develop-ment also has a Parkinsonian component: studies and analyses are being pursued almost for their own sake, with very little operational equipment in actual readiness in the hands of the armed forces.[27] Development delays are multiplied by the many mutually contingent decisions to the point where the equipment, if ever procured in quantity, is on the verge of obsolescence by the time it has reached operational inventory status.

The psychological impact of this evolutionary process, insofar as it affects military personnel, should not be overlooked. Instead of being a field to exercise the most highly praised martial virtues, courage, heroism, and sacrifice, even enlisted personnel must nowadays operate in a sophisticated equipment environment against an unseen mysterious enemy. Victory may belong to the side whose

[8] "Conventional" is used in contrast to "nuclear" weaponry.

computer happens to be still operational, just as death and defeat results from completely unknown actions of faceless opponents. Still as a consequence of the equipment sophistication and the associated advanced professional training requirements, it is likely that future "enlisted" personnel will be less prone to offer blind obedience and to accept personal sacrifices than his (or her) predecessors.[9]

* * *

If large-scale wars, possibly but not necessarily involving employment of nuclear weapons against the superpowers, are accepted as being unattractive options from the Western viewpoint, one may argue that the more traditional use of military forces, i.e., the securing of markets and sources of raw materials by means of threats against, or actual possession of, overseas territories, might still appear as advantageous, and hence legitimate. The (possibly) regrettable truth is that, for reasons just as fundamental as those affecting central nuclear conflicts, the modern equivalents of colonial wars are doomed categorically to failure and should not be considered seriously as part of the Western long-term strategy.

We live in a world in which, for better or worse, the major ideological competitors of the West have acquired political legitimacy, substantial power bases, and (in the case of the Soviet Union) first-class modern military capability. In consequence of the technical nature of advanced weaponry, significant military capability can be "transfused" from major powers to minor client nations (and even to non-sovereign groups) with considerable effectiveness in relatively short time. This, in practical terms, means that the major military advantage of the West, which is the ability to bring to bear its military power overseas against relatively weaker and unprepared opponents, is irretrievably

---

[9] (Note added in proof) It is to be hoped that the low motivation, lack of discipline, and below-average IQ of Army recruits, currently attributed to the Volunteer Army concept, will prove to be temporary deficiencies.

lost. Whenever the Soviets (and, eventually, other hostile groups) choose to oppose us in a hypothetical overseas venture, they can do so indirectly by simply transfusing the appropriate type and level of military equipment, together with advisors, training, and logistics support.[28] In the days of unquestioned strategic Western supremacy, such action might have been deterred even prior to its inception. Even short of successful deterrence, the unquestioned Western advantage in sea power and the notorious logistical insufficiency of the Soviets would have made such operation risky from their viewpoint.

None of these restraining influences can be counted upon at the present and in the foreseeable future. The Western strategic deterrent capability has been allowed to weaken to the point that it is no longer credible in this context; the trend of Western vs. competing navies raises questions about our continuing ability to control the sea if more than one theater becomes critical at the same time; long-range airlift, including the use of dual-purpose civilian transport aircraft, becomes increasingly available to our opponents and, most importantly, the relatively small tonnage of modern weaponry can quite seriously change the nature of military confrontations.[10] In more general terms, if a coalition of Western nations elects to respond to a crisis with the threat of military intervention in some Third World nation typically devoid of significant military capability, and if Soviet Russia, or in the longer term future even some other Third World group, elects to equip the target country with modern military weapons, the task of the Western protagonists becomes rather burdensome and the outcome problematic.

The foregoing argues mostly the *reactive* mode, i.e., how Western military initiatives may be countered by relatively

[10] Egypt in 1970-71 in a matter of a few months acquired a modern air defense system that, to say the least, has created some new problems for the Israeli Air Force. When common borders exist, e.g., between North Korea and the PRC, or when unimpeded shipping is available, e.g., between Vladivostok and Hanoi in 1972-1973, the transfer of medium and heavy equipment becomes also possible.

weak nations aided by timely transfusion of military equipment and combat readiness. The *active* mode of this same principle, i.e., how to modify the status quo in a manner hostile to Western interest, is also possible and is at the present juncture gaining in popularity and in frequency of occurrence. By providing equipment, agitation, propaganda; by contributing "advisors" and even combatants of proxy nations[11] the Soviets may encourage and support any splinter group (fashionably called "liberation front") or guerrilla band willing to initiate military or paramilitary action hostile to the West, which can last indefinitely unless suppressed by rather violent and nondemocratic actions. Thus, the outcome in insurgency and guerrilla-type warfare is, by its very nature, biased against governmental authority which attempts to respect the laws of humane warfare or which expects to control the population, while respecting democratic principles.

In this perspective, any policy based on Western hegemony in a given sphere of influence, such as the early 19th century Monroe Doctrine, may present serious implementation difficulties, unless it has steadfast and effective support of the "beneficiary" governments *and* populations. Any publicized political and military disengagement of the West from areas which are arbitrarily defined as "nonvital" conversely contributes a "hunting license" for direct or indirect intervention by our adversaries.[12] The difficulties experienced by the U.S. in the Caribbean (Cuba, Nicaragua, and, as of the time of writing, El Salvador) in trying to stem the spread of "leftist" or "nationalistic" trends illustrates the problems of maintaining the political status quo in an area and at a period when borders are essentially permeable to ideas and war-like equipment, as well as to dedicated cadres. The recent decade of Middle East history may equally serve as an illustration, a drastic one at that, of

11   The role of Cuban soldiers in Angola, Congo, Mozambique, and Ethiopia should be seen as foreshadowing this type operation.

12   Perhaps the most vivid illustration of what has happened to Western power and global influence is obtained by juxtaposition of two texts: The Monroe Doctrine (1804) and the Nixon Doctrine (1969).

what is liable to happen when the West (in this case Great Britain) elects to voluntarily relinquish its responsibilities "East of Suez."

* * *

The experience of the last twenty years and the projections for the next few decades suggest that the Soviets will oppose by all available means Western attempts to preserve or to restitute the integrity of the global economic system as it existed until about 1965. They will also support and exploit any perturbation if it can conceivably lead to a decrease in prestige or influence of the West. The threat, and the direct or indirect use, of military power is clearly and explicitly part of pursuing this policy.[13] The Soviet attitudes toward the deliberate use of military force in furthering their national objectives should be examined at this point.

The communist doctrine, as promulgated by the Soviet Union, holds that the capitalist nations and their fascist offsprings are committed to the destruction of communism. It accepts as axiomatic that the bastion of international communism, Soviet Russia, must develop to the fullest its defensive military power in order to counter the explicit threat posed by capitalism. Still according to the doctrine, the counter-revolutionary machinations in the "friendly socialist nations" must be thwarted by the military power of Soviet Russia, if national forces are found unequal to that task. More pragmatically, but in keeping with the centuries-old tradition of Imperial Russia, the maintenance of law and order within the boundaries of the USSR is also routinely assigned to the military. Even with the unalterable certainty that the capitalist-bourgeois regimes will eventually fall through their own internal contradictions, it

---

[13] The significant exception appears to come to the fore when a trend or an incident seem to threaten the internal cohesion of the Soviet Union itself. The Islamic "Renaissance" currently experienced in the Middle East is a case in point.

is understood that the "socialist correlation of forces" comprising political, social, economic and (most emphatically) military actions will greatly accelerate the process.   In the meantime, the struggle of the exploited peoples for national liberation must be helped by all possible means, including military equipment, training, logistics support and, when necessary, "comradely" participation in actual military operations. Particular vigilance is required in order to preclude the capitalist governments from unleashing nuclear war on the world, as a final act of desperation in the death-throes of their regime. For this reason the Soviet nuclear forces must be ready at all times to overcome and to destroy the nuclear weapons of the U.S. and its Allies.

The Soviet doctrine thus fully supports, and as a matter of fact requires, priority for military preparedness over most other national objectives of the USSR. While subjected to the control of the Politburo, the senior Soviet military officers are essentially part of the government; if the paramount importance of military power is ever questioned, this has not been detected in recent times by Western analysts. As evidence of this national commitment, it is generally recognized that the Soviets have essentially achieved local superiority in conventional air, land, and naval combat capability just about everywhere along their boundaries; it is also taken for granted that the Soviets currently have, or at least will very soon acquire, the strategic nuclear weapon parity or even supremacy which would essentially negate the U.S. deterrent doctrine, except perhaps as it relates to direct, massive attacks against the continental U.S. itself.   At variance with the U.S., the Soviets have also chosen to invest in an elaborate and reputedly competent air-defense network (possibly adaptable to the antiballistic missile defense mission) and, more recently, in a long-term program of industrial and civil defense, clearly and admittedly aimed at nuclear war-fighting and subsequent recovery capability.

According to Western military experts, the weak point of the Soviet military posture is the relative lack of surplus capability in the production base coupled with the weakness of the transportation and logistics system. Even in peacetime the Soviet agricultural production is frequently falling short of its assigned production targets; as to the industrial production capacity and the adequacy of the transportation/logistics system, it is quite probable that the stresses of major long-protracted wars would rapidly become unbearable. Two considerations mitigate this perception, though: in a centrally controlled, authoritarian society the standards of performance used in the West have a tendency to evaporate[14]; furthermore the Soviets do definitely not plan to refight a major non-nuclear war in which the industrial resources of the Western nations could be effectively mobilized.

If the burden imposed by the massive military program, which has been on-going for the past twenty years, stresses the Soviet economy to its limits of endurance, related evidence is not clearly available to the West. There is no meaningful public debate on defense expenditure and planning levels; the public seems to be enured to unilateral government decisions, invariably emphasizing long-term investment goals, i.e., heavy industries, infrastructure, and the military, at the expense of consumer amenities. The Soviet planning and budgeting process offers ample opportunity to disguise, or to publicize, the specifics of defense-related outlays;[15] and since the government has complete control of the economy (prices, wages, taxes, saving levels, interest rates, etc.), the funding available for military programs is to a large extent discretionary. If there are any criticisms by overzealous news reports or hostile inquiries by elected representatives, these do not seem to be given much publicity.

[14] The almost unimpaired production capability of Germany during World War II should be recalled in this regard, which in fact *increased* during all of 1944, the year of almost complete blockade and intensive strategic bombardment by the Allies.

[15] A public debate on the desirability of committing the enhanced radiation weapon ("neutron bomb") to full-scale production, such as took place in the U.S. in 1978, would be quite unusual, not to say unthinkable, in the USSR.

Mostly for these reasons the cost/effectiveness of military procurement does not, by far, have the same mythically venerable connotation as it has acquired in the West. Even though labor shortages may appear in certain critical areas, the Soviets would not conceive of scrapping an obsolescent weapon system just because another, more efficient, weapon is on the verge of entering the operational inventory. The technological aspects of military preparedness have equally ceased to be critical in the sense of limiting Soviet developments; the military can and does command the very best of skills and talents to be found anywhere within the Soviet bloc; it has absolute priority on materials, equipment and facilities. Red Army planners also have apparently found the right balance between operational suitability, serviceability by relatively nonspecialized personnel, and the military effectiveness when matched against Western-supplied equipment—especially in the hand of allies and proxies whose training might be of relatively recent vintage. In terms of battlefield tactics, the Soviets add to their traditional principles of massive application of concentrated resources and exploitation of surprise the new emphasis on deliberate deception and disruption of the adversary's command and control structure.

With all these apparent advantages, one should not be surprised to observe that the Soviet government considers the use and the threat of force as one of its essential foreign policy tools to be used both in the active (offensive) and passive (defensive) modes. In relation to their allies, neutrals, and potential enemies, the presence of rapidly deployable military strength-in-being is never far below the surface. Their often demonstrated readiness, their apparent ability (in the past) to take rapid action, their reputation for ruthlessness—these all, in combination, form a rather powerful set of adjuncts to any influence they choose to exert, or any negotiating position they elect to support.

This doctrinal attitude toward the role of military power is in sharp contrast to that prevailing in the West. The actual propensity of the Soviets to *directly* participate in wars with major Western nations on the opposite side is nonetheless subject to many ambivalences. Some of these clearly are at work at the present time, some others are predicted with a fair measure of confidence for the future. What is more, other totalitarian societies may well be influenced by these same factors in the longer term perspective.

The Soviets have exhibited in the recent past years a remarkable measure of caution. Some of this may be seen as a deliberate intent of presenting a reasonable and peace-loving image to the world, but many observers attribute this caution to the desire of avoiding any serious military risk to the Soviet Union, whenever other avenues associated with reasonable promise remain open.[16] Equally on the side of restraint is the inherent distrust of the ruling party elite toward all other segments of their society: the mass of hard-pressed working classes; the new generation of white-collar intelligentsia, containing a high proportion of disaffected if not downright dissident elements; some of the European satellites less than a generation away from rather bloody repression by the Soviet military; those within the Soviet "socialist" republics whose racial origins and religious backgrounds would predispose them to regard the dominant White Russian minority as the oppressors. In the latter case, the pressures of demography and the Islamic renaissance are understandably eyed by the Soviet futurologist with no small measure of apprehension.

Balancing these restraining tendencies, one must also consider the internal and external trends which would press for relatively rapid rise in the willingness of military risk-taking. The Soviet military estabishment, having received the lion's share of disposable investment of the nation for more than two decades by now, may be pressured to show some tangible payoff in terms of economic improvement, if (as certain Western analysts see it) the Soviet economy runs into resource-related constraints in the mid or late 1980's.[29] Perhaps the most potent element which could conceivably impose, or accelerate, a Soviet timetable for the deliberate employment of military power is the global cultural and

16 (Note added in proof) The important exception is the forceful application of the "Brezhnev doctrine," first formulated explicitly in 1968 for Czechoslovakia. As currently observed in Afghanistan and Poland, it can receive quite elastic interpretations, especially when the U.S. is embroiled elsewhere.

economic evolution, strongly suggested by the preceding chapters. Information is rather abundantly transmitted around the world; the ideological opponents (including the West) just may learn the value of effective propaganda. The emerging countries learn to compare, through numerous (albeit indirect) exposures, the relative blessings of various political systems; it is not obvious (perhaps not even in the Soviet mind) that such nations, if given a free choice, would willingly espouse any socialist doctrine, let alone *their* brand of communism. The acquisition of conventional armaments, relatively powerful in a local context, with the interested encouragement of both East and West, and mostly the uncontrollable longer term spread of nuclear weaponry, simply make the world more difficult to influence and less susceptible to ideological control. The Soviets may view this evolution with dismay mingled with alarm—especially if Red China continues to grow (with conspicuous help from the West) into a fully competitive participant.

The military perspective as seen by the Soviets is thus quite different from the one observed from the Western vantage point. They are now locally stronger than the West in many places (except, for the time being, in the Western hemisphere), they can trust their military and logistics for relatively short wars, and they may soon believe that their nuclear supremacy may deprive the West of its ability to control escalation. Nothing in their political doctrine or tradition would oppose the use of force, except their desire to avoid unnecessary risks. While a number of factors appear to urge further and increasing caution, there are others which could well be interpreted as favoring delib- erate future attempts to exploit their perceived military superiority. Perhaps the most credible driving force toward military confrontation could be a clearly observable hardening of Western political attitudes, supported by massive, concerted and fast-paced efforts to recover their military superiority in areas judged vital from the Soviet viewpoint. It is far more likely, however, that the Soviets will continue their relatively cautious policy, with the rates of investment in military capability carefully maintained at a level sufficient to assure a satisfactory edge against the West, but mostly against Red China and its allies. In the meantime, they will take strong interest in the political and economic travails of the industrialized world, ready to exacerbate and exploit blunders, opportunities, and accidents.

* * *

Local and regional wars, pursued mostly for reasons of influence, prestige, or economic interests, can convincingly be shown to offer few promises and even fewer long-term benefits to the West; this should be sufficient to rank them rather low on any preference-ordered list of strategy options. In addition though, and most emphatically in view of the observations made about large-scale wars, the inherent connection between the two—local/regional conflicts and large-scale, possibly nuclear, wars—must be taken into account. There are ample reasons to believe that no rational person would deliberately initiate (nuclear) war, but this belief presupposes a set of shared values and a common basis for the definition of "rationality." Unfortunately, wars are not initiated or escalated on a rational basis; once started, they acquire a logic and a momentum of their own and the escalation of small, or even insignificant, conflicts into progressively larger confrontations, culminating in central nuclear war between superpowers is an ominous threat and an ever-present possibility. As long as the strategic (nuclear) balance was clearly and admittedly in favor of the West, one could (with more or less conviction) speak of "crisis management" or "escalation control"; the reader should be persuaded by now that the West has deep-rooted interests in avoiding nuclear wars, while protecting the status quo insofar as possible. Once the relative strategic postures have been explicitly and emotionally recognized as "balanced" or "equipoised" (with the obvious and significant margins of error associated with any military estimate, let alone with one which has, mercifully, never been tested in human experience) no one side "controls" escalation, none "manages" the crisis and the world may slide into nuclear war through miscalculation or misperception. If the strategic balance is allowed to further shift in the direction favoring the Soviets, the West will be perfectly safe from any temptation of initiating or encouraging local/regional conflicts: at the first glimpse of such proclivity, our adversaries will, most assuredly and most forcefully, set us back on the path of righteousness.

* * *

A formidable endeavor, such as recreating and maintaining the Western military might in a state of permanent readiness, is bound to have deep social repercussions. Citizens and their elected representatives alike observe the manifestations, and ponder the implications, of the massive military expenditures which, by 1980 and measured in that year's inflated currency, will have reached the staggering yearly total of a quarter-trillion dollars for the group of Western nations.

In societies where freedom of expression and pluralistic formulation of public policy are the norm, uneasiness about the unbounded power or dissatisfaction with the suspected weakness of a nation's military posture are bound to arise; in periods of relative quiescence the forces opposing defense spending appear to exhibit inherent tendencies of self-reinforcement.   An essential new trend has surfaced in recent years: *the nation's perceptions in regard to defense preparedness can be systematically and effectively manipulated by interests opposed to those of the West.* The free societies offer ample opportunity to adversaries for influencing their public decisions; the pertinent information is freely available or can be obtained by relatively unsophisticated intelligence efforts, when not deliberately "leaked" by private (or for that matter, public) interest groups. From within centrally controlled societies, by way of contrast, meaningful information is by far more difficult to acquire; the messages potentially affecting Western behavior can be fabricated and supported by observable (albeit deceptive) governmental actions.[17] An intelligent and

17   Appendix - Note F.

perceptive adversary, having knowledge of the mechanisms and the substance involved in Western policy formulation, having furthermore little or no internal checks and balances on the conduct of its foreign affairs *including military decisions* can, to a considerable extent, manipulate the perception of external challenges by the individual Western nations and by their alliances. There are ample reasons to suspect that such "strategic manipulation" is in fact being deliberately pursued with increasing sophistication and effectiveness. If so, this may constitute one of the major causes of the continuing erosion in Western military power. Unless the impacts of this type of interaction are carefully explored and deliberately countered, the future military options available to the West will become even much less attractive than they appear to be at the present time.

***

In summary, owing to a combination of political, societal, and technological causes, the Western perspective on the use of military power, as an effective means to further assertive foreign policy objectives, has radically changed in the past few decades. For the first time in history, the Western nations, as a group, face the Soviet Union, a dedicated adversary of the first magnitude, who challenges their political and social concepts and who has the will and the military power to back up this challenge.

The Soviets emphatically refuse to share the Western view on the future of warfare; well to the contrary, their tradition and doctrine explicitly support the use of force to spread their ideology and to enforce their domination.

Major wars became unthinkable for the Western democracies as the nuclear balance reached a state of rough equivalence between the superpowers. By way of contrast, the Soviets prepare and advertise their warfighting capability, including nuclear wars over their own homeland. The principle of strategic deterrence, so long the cornerstone of the U.S. strategy vis-a-vis the Soviets, is now verging on obsolescence.

The West also has in effect lost its "crisis management" or "escalation control" capability; and local or regional wars can no longer be undertaken with the hope of termination under favorable conditions.   The Soviet Union, and eventually other hostile groups, can render overseas ventures even less rewarding by transfusing military training and equipment to our opponents, by supporting combat strength through the use of proxies, and by encouraging nonconventional forms of combat (such as terrorism and insurgency), in which the conventional military forces of the West cannot be used to advantage.

The Western morale, national temper, the will to fight, and the strong, sustained commitment to superior military posture can, under present and foreseeable conditions, be manipulated to a significant extent by those with opposing interests.

In all of the foregoing, three key elements constitute a common theme. The changes observed are global in scope; they can probably not be reversed in less than a decade or so; the resulting perceptions impact the two opposed worlds asymmetrically. The doctrinal views on warfare, the preparedness for nuclear warfighting, the attitudes and expectations regarding local and regional wars, the manipulative interactions with the opponents' public policy formulation—all these exhibit the characteristics of global scope, quasi-irreversibility in the near term, and differential impact on Western, as compared to Soviet, societies.

Military undertakings appear increasingly unpalatable and unrewarding to the West; many find it difficult to assert that the same will reliably hold true for its opponents.

Part 3
ACCELERATING THE ADAPTATION PROCESS

CHAPTER IX
The Modified Premises

The reader, in the wake of the preceding chapters, should by now be satisfied and convinced that a number of large-scale, secular changes in our social environment are in the process of taking place. They have been shown to contribute, independently or in combination, to the stresses and to the growing atmosphere of *malaise* which started us on our quest in the first place. We are now in position to first define, then to examine in some detail, the changes in the major premises of our domestic and foreign policy formulation that could possibly improve our societal "response" or, to use the words of our original hypothesis, could *accelerate the process of cultural adaptation* to the shifts in the environment. This chapter offers a road map to the remainder of our discussion.

The new or modified premises are stated boldly and assertively, mostly to entice the reader's further efforts by challenging or corroborating his or her currently held views. More detailed discussions of the rationale behind the assertions are prudently relegated to subsequent chapters, but a few preliminary comments are in order, though, right at this point.

The word *premise* is used advisedly. Both in the context of currently proclaimed or implied premises and the possibly desirable new or modified ones, it refers to axiomatic assertions, serving as basis for arguments leading to policy formulation. The consistency and the effectiveness of the resulting policies are the major, if not the sole, acceptability criteria for such premises.

Attempts aimed at defining modified premises for policy formulation should most emphatically not be likened to, or confused with, efforts to forecast the future, or to use a more fashionable term, the efforts of "futurologists." The desirability of modified premises is derived, one may hope, by following a reasonably plausible rationale (as distinguished from rigorous logic) based on observable trends; the resulting premises are offered mostly for further argument and, if found satisfactory upon far more detailed scrutiny, for implementation. Forecasting, by way of contrast, attempts to predict future trends, singly or in combination, based on current trends with the help of more or less explicit theories. The reasons for disclaiming kinship of the present study with futurology are as simple as they are obvious: Predicting the future is hazardous business, fraught with serious difficulties.[30] It has in recent years, perhaps undeservedly, fallen into somewhat of a disrepute, owing to some rather glaring failures[1] and even to some instances of subservience to the client's mental predispositions.[2] Even more important is the fact that, while the present study (in common with forecasting) draws heavily on the examination of past trends, it carefully refrains from applying the same to predicting the future.

On the other hand the search for adequate premises for long-term policy formulation was strongly suggested and encouraged by some of the phenomenally successful

---

[1] Vannevar Bush, one of the most respected scientist-statesman of his period, flatly ruled out in the mid-1940's ICBM's as practical or promising elements of future warfare. In the late 1940's the large-scale digital computers envisioned by Von Newmann and his associates were peremptorily rejected by hardware experts on the grounds of complexity and unreliability,[31] the advent of ultrareliable, microminiaturized and mass-produced electronic components was not anticipated, even by the giants dominating the vacuum-tube industry, as late as the early 1950's. Closer to our time, when presumably the technique of forecasting had somewhat progressed, in 1972 one of the most brilliant and respected group of U.S. futurologists in a book describing the 1970-1990 period, left essentially undiscussed problems such as oil cartels, resource scarcity, energy crisis, cumulative gold flow to the Middle East, and Islamic Renaissance.[32] Many other instances could be mentioned, mostly to illustrate the difficulty of truly meaningful forecasts, even for the relatively near term.

[2] Appendix - Note G.

examples of the past. In most of these, explicitly formulated premises have led to long-term policy objectives, which in turn were supported by felicitous confluence or concatenation of many favorable motivational, economic, and technical factors. Examples readily come to mind: The origins of British naval supremacy during the reigns of King Henry the VIIIth and then mostly of Queen Elizabeth; the land-grant policy of the 19th century U.S. to accelerate the development of railroads and state colleges; closer to our time, the deliberate policy of Japan to invest in growth sectors of the advanced industrial economies (e.g., textiles, electronics, optics, shipbuilding) and then just as deliberately to upgrade and reorient their industrial skill spectrum whenever a given sector ceases to exhibit the characteristics of vigorous and profitable growth. Perhaps the most fascinating example is the confrontation of the Western world with the 15th century predecessors of the Arab participants of the OPEC.[3]

It is amply proven by historical examples that long-term policy objectives, based on sound premises and sustained for several generations, often result in major competitive advantages for the nation or group of nations which has chosen to underwrite the commitment. This was quite consistently true even at periods when the formulation of premises and policies was mostly a matter of faith or belief in some mythical destiny, rather than the result of (allegedly) rational cost/benefit analysis. A measure of faith and emotional commitment among the leaders may still be essential elements of success, but we should nonetheless find additional reassurance in the pros and cons discussion in keeping with more conventional logic.

* * *

Without further apologies or justifications, here then are the new, or at least modified, premises which could be of help in the process of formulating our future domestic and foreign policies in view of alleviating the observed causes of stresses and instability--in short, which could possibly accelerate the process of cultural adaptation.

3 Appendix - Note H.

Chapter X deals with the domestic structural problems of "Western" democratic societies[4] at the national level. All these individual nations have constitutions or established traditions to protect the citizens' rights and privileges (mostly against encroachment by governmental authority); the problems of societal groups are, in general, not explicitly given the same degree of emphasis. The U.S. Constitution and its Amendments appear to be mostly concerned with the protection of citizens from the states and with the relationships between the (theoretically sovereign) states and Federal Government. Protection and well-being of the nation as a whole are being provided by clauses related to "common defense," to the meticulous attention given to "checks and balances" among the three branches of Government; to the promotion of foreign trade, "science and useful arts." It is implicitly assumed that once these *sine qua non* conditions are provided, the nation will be prosperous, will enjoy the fruits of liberty, and thus the coherence and stability of society are essentially assured.[5] The modified premise, perhaps more appropriate to modern conditions might read as follows:

FIRST PREMISE: THE MAINTENANCE OF SOCIETAL COHESION AND STABILITY IS AN EXPLICIT GOAL OF EACH "WESTERN" NATION, TO BE PURSUED ON PAR WITH THE PROTECTION OF THE INDIVIDUAL CITIZENS' RIGHTS.

Chapter XI extends the same thought to the domain of foreign policy. The "Western" nations, through common cultural and political inheritance, do possess the intrinsic capability to act in concert, provided that the appropriate decision framework and action mechanisms be available. It is reasonable to suggest that the interest of individual nations in this group may be better served in the long term if such framework and mechanism for concerted action are

---

[4]   The terms "Western" or "Western-style" societies will continue to be used more as a descriptor of broad political philosophy than carrying historical, racial, or geographical connotations.

[5]   Deliberate (slow and ponderous) procedures are provided for peaceful means of changing the Constitution and other aspects of Government.

planned long in advance of specific need. This leads us to the

SECOND PREMISE: THE LONG-TERM INTERESTS SHARED IN COMMON BY THE WESTERN-TYPE NATIONS DO (AND WILL) BY FAR OUTWEIGH THE CAUSES OF COMPETITION AND CONFLICT AMONGST THEM. THEIR GOVERNMENTS SHOULD PURSUE THE MAINTENANCE OF THE COHESION AND STABILITY OF THE GROUP OF THESE NATIONS AS AN ESSENTIAL ADJUNCT TO THEIR DOMESTIC-ORIENTED NATIONAL GOALS.

Pending more detailed discussion, we should most emphatically observe that the implementation mechanisms should enable the accretion of the "Western" group[6], i.e., allow other nations to join, with various degrees of propinquity or formality, as and when the evolution of their own international perspective prompts them to do so. The important and growing population foci, India, Brazil, Mexico, and Indonesia, at various stages of political and economic evolution, are the specific instances where this may occur within the next two decades.

Both the first and the second premise lose their plausibility if the economic basis of the "Western" way of life is seriously weakened. The very expressions "industrial" and "post-industrial" connote efficient, broad-based production, distribution, and consumption; with the accumulated capital being reinvested in growth, and more to our point here, in spreading social benefits to an increasing number of (economically) less efficient participants. In order to support and stabilize this process at a world-wide scale, Chapter XII addresses the

THIRD PREMISE: THE LONG-TERM STABILITY AND WELL-BEING OF THE WESTERN NATIONS, INDIVIDUALLY AND COLLECTIVELY, ARE CONTINGENT UPON THE DEVELOPMENT AND IMPLEMENTATION OF A CULTURAL AND ECONOMIC INTERACTION FRAMEWORK WITH THE LESS DEVELOPED NATIONS, ATTRACTIVE AND ACCEPTABLE TO ALL THE PARTICIPANTS.

6 See footnote 16, Chapter VII.

Note that this phrasing specifically covers the potential resource suppliers as well as the "have-not"[7] nations, irrespective to their then current political alignment. The premise, as stated, emphasizes the word "attractive" to cover all the cultural and ideological factors likely to color the respective value judgments.

Chapter XIII is dedicated to the proposition that decisions affecting human group governance, of necessity and almost by definition, involve coercion. The implementation of policies derived from the first three premises (which of course must remain in reasonable consonance with the many currently ongoing and projected processes) may require periods measured in decades, if not in generations. Many imperfections in the course of future evolution, many differences in value judgments among the participants, many attempts by its adversaries will often, perhaps critically, endanger the long-term outcome.[8] Even assuming the relatively crisis-free evolutionary period, experience of all human organizations proves that even the steady-state operation requires the presence of carefully controlled, but effective, coercive power to subdue unlawful manifestations of aberrance. This leads to the

FOURTH PREMISE: THE AVAILABILITY OF EFFECTIVELY USABLE MILITARY POWER TO THE GROUP OF WESTERN NATIONS IS AN ABSOLUTE PREREQUISITE TO THE ESTABLISHMENT AND PRESERVATION OF STABLE POLITICAL AND ECONOMIC STRUCTURES COMPATIBLE WITH THEIR COMMON IDEOLOGY, AT THE NATIONAL AND GLOBAL LEVELS.

---

[7] This term is reproduced here mostly to express my violent revulsion against its use. The nations thus described have what is potentially the essential wealth of the Western nations: individuals who all, at different times and in various combinations, eat, fight, work, learn, and reproduce. The past pejoratives applied in the domestic context ("serfs," "souls," "villains," "sans-culottes") have long ago been sublimated into "masses," "proletariat," or simply "workers." To ignore this aspect of cultural evolution at the international level would, in all probability, be a catastrophic mistake.

*

Chapter XIV is devoted to the discussion of the overlaps and mutual dependency among the implementation aspects of the policies derived from the preceding premises. Just as the description of the environmental changes (Part 2) had to proceed along somewhat arbitrary and overlapping categories, we find here again that the premises can not be independently   discussed,   and   most   certainly   not implemented in isolation. This observation, elicited by the purely mechanistic attempt at organizing the discussion, may lead to one of the major conclusions of this work.

The perceptive reader will also notice that the premises, as stated, do not exhibit complete one-to-one correspondence to the individual components of the environmental changes discussed earlier. The reasons are matters of logic, cause-to-effect relationships, and hierarchy. I have found it more appropriate to discuss the information environment first, as contributory to the causes of decrease in societal cohesion as well as to the destabilization of the global economic system. The resource/ecology topic was treated separately as an illustration of an important stimulus/constraint coupled with often hasty and unwise response by society. Both of these components (information environment and resource/ecology) were found to be logically associated with the first three premises.   Their respective impacts are therefore discussed in the corresponding chapters following.

It may be wise at this point to remind both the writer and the reader of the purposes and limitations of our inquiry. As stated in the opening words of this section, we wish "... to examine in some detail the changes in the major premises... that could accelerate the process of cultural adaptations..." It would be quite preposterous, in a work of limited scope

---

8   It is useful to recall that even the "more perfect Union" purposed by the U.S. Constitution had a number of crucial tests of stability where the force of arms was the decisive factor in determining the outcome. The Civil War is, of course, the best known because of its magnitude, but the U.S. Army has been called upon to support law and order as recently as in 1957 (Little Rock, Ark.), and in 1962 (U. of Mississippi), when Federal authority was challenged by the state governments.

such as the present one, to suggest that we plan to discuss a complete blueprint of how democratic societies ought to be organized and governed. Many essential aspects, such as inflation, the stability of currencies, or the shift of the demographic composition of post-industrial societies toward the older age groups, are barely being touched upon. I hope nonetheless that some of the suggested changes in policy objectives will be found sufficiently intriguing to prompt further exploration.

CHAPTER X
National Cohesion and Stability

FIRST PREMISE: THE MAINTENANCE OF SOCIETAL COHESION AND STABILITY IS AN EXPLICIT GOAL OF EACH "WESTERN" NATION, TO BE PURSUED ON PAR WITH THE PROTECTION OF THE INDIVIDUAL CITIZEN'S RIGHTS.

Owing to their historical origins, the constitutions of all Western nations are replete with principles, axioms, and "self-evident truths" aimed at protecting the individual. The religious beliefs, the physical security, the property of citizens; their freedom of speech, assembly; their consent to government, both in form and in substance--are all specifically surrounded by explicit guarantees.[1] Whether explicitly stated or only implied, the Government's responsibility for the protection and stability of society as a whole is just as clearly understood. In the U.S. Constitution the concern about external threats ("common defense"), about economic growth ("regulate commerce...coin money and regulate the value thereof...promote the progress of sc˙ence and useful arts"), about infrastructure development ("...post offices and post roads...") is evident; in point of fact, Article IV even guarantees the "republican form of government" to every state in the Union as well as protection against domestic violence. With all its exquisitely detailed precautions to ensure that the Government is truly representative and

---

[1] The U.S. Declaration of Independence and Constitution, with the Bill of Rights and subsequent Amendments, are taken as an example. As in previous chapters, the subsequent evolution is discussed in terms mostly relevant to the U.S., based on the belief that they are applicable *mutatis mutandis* to other Western nations as well.

responsive to the will of the electorate, the Constitution reflects the Founding Fathers' deeply held conviction that given the large measure of individual protection and freedom, given the (minimal) explicit governmental role in protecting the nation as a whole and in promoting common welfare, and given the checks and balances among Government branches as well as continuous and effective checks by the voter-citizens, the cohesion and stability of the society is *ipso facto* ensured. As a further precaution, ponderous but peaceful means were provided to change even the Constitution when such changes are judged necessary.[2]

Over the past two centuries many adjustments have taken place. The most important ones have in fact been sanctioned by Constitutional Amendments (e.g., abolition of slavery, levying of income taxes, establishment of universal suffrage); others were enacted by Congress and upheld by the Courts to become the "law of the land." In this second category many attempts were made in recent decades to eliminate or to mitigate the societal stresses as they became apparent .[3] This is not the proper place to examine all these attempts, but three significant aspects should deserve a closer look:

• As the volume of information reaching the individual increases beyond his individual ability to select, to assimilate, and to use effectively such information for his benefit, the effective control of the people over Government actions becomes a fallacy.

---

[2]   The framers of the U.S. Constitution intuitively provided explicit mechanisms for sociopolitical adaptation, about a century before evolution was seriously considered as an element of social sciences.

[3]   The 1929-1933 "Great Depression" mandated the Federal Government to insure bank deposits, to provide for equitable labor relations and Social Security; in the 1960's legislation was passed to make the Government responsible for "full employment" even at the expense of becoming "employer of last resort"; Medicare and Medicaid have been added to Social Security; equal opportunity, environmental protection, occupational safety—all have become (perhaps erroneously) proper concerns of the Federal Government.

- As the volume and complexity of social and economic interactions increase, the scale of unintended, but apparently unavoidable, fluctuations creates excessive burden on broad classes of individuals, for all practical purposes independent of their own actions. In most of these instances, those experiencing hardships turn to the Government for remedy.

- The preponderant emphasis on relatively near-term concerns often precludes appropriate consideration of the longer term consequences of regulatory and judicial decisions.[4]

These three observations form the basis of the rationale offered in support of the first premise; i.e., that public authorities, in concert with individuals and private organizations, should do more to ensure that society's longer term cohesion and stability are fostered and protected.

The policies derived from the first premise would then fall logically in three categories: those which are concerned with the information reaching individual citizens; those which are aimed at the economic stability and the incentives to productive activities; and those which promote the public understanding of longer term issues, ensuring thereby a better balance in the decision processes between the near- and long-term pressures.

There is nothing categorically new or different in the thought that democratic governments should be constructively concerned with society's cohesion and stability. In point of fact, even the most totalitarian governments place their preoccupation with the perpetuation of the "state"

---

[4]   The 1974 decision of the U.S. Supreme Court to mandate Congressional redistricting on the "one man, one vote" principle properly reflects its concern with Constitutional principles. It also may have biased the Legislative branch even further in the direction of near-term emphasis by giving additional voting power to the urban poor having little or no interest in the longer term concerns of society. Some over-emphasis in the socially oriented "reordering of priorities" may find its cause here.

(system of government) ahead of any other consideration in their priorities.   The implementation policies discussed below should be seen as embodying, or even emphasizing, the basic democratic ideal which recognizes that no single individual, group, or class can lay claim to the revealed truth in matters of government; that none of these has historically exhibited a degree of wisdom or unselfishness which would justify its dominant influence; and that the (well-informed) citizen's free consent is still the essential test of his satisfaction with the prevailing institutions.

Let us now examine some of the implementation policies that could be logically derived from the first premise.

**

The right of individuals to select, to control, and to use the information reaching them should be more effectively protected.

The integrity of the political process, as well as the usefulness of individual choices, are conditional upon the adequacy of the available information base and the ability of each of us to mentally "process" this information. Everyday's experience, formal education, advice, motivation, or even manipulation by others are the tools and the references for this processing.   Our present social behavior stresses through a multitude of means the improvement of our individual processing or decision-making ability, but places a somewhat distorted emphasis on the availability of input information.   It is implicitly assumed that an educated or well-adjusted individual (i.e., one who uses the right criteria for decision making) will automatically filter out the useful portions of the information stream, rejecting those which are harmful, untrue, or misleading.   Most of our legislative and behavioral tenets stress the free availability of information.   In view of the technical advances in information generation and mass transmission,[5] this emphasis might

[5]   Chapter IV.

be grossly misplaced. The volume of information which reaches an individual has become so large that his or her ability to use it effectively may have become impaired. Hence the suggestion that it is time for us to shift our attention to the *quality* and *nature* of the information flow rather than just expressing concern about its continuing availability.

While in the early history of Western democracies the emphasis was on the protection of the sources, in more recent times attention has shifted to the protection of the recipients. In the days of absolute monarchy, the central problems of those interested in fostering individual freedom and increased individual participation in the affairs of government were the means for generating and disseminating their ideas, as well as the protection against arbitrary disfavor or retaliation by those interested in protecting the *status quo.* In more recent times the freedom of thought or of speech is firmly rooted in the democratic traditions; as to the means of dissemination, not only are they available in great abundance, they are also technically and economically accessible to anyone seriously inclined to use them. On the other hand, the recipients, as a group, readership, audience, or "viewership" are for all practical purposes defenseless against relentless competition by all media for their attention.[6]

There are many historical instances where the constitutionally guaranteed freedom of speech or freedom of press have been abridged in order to protect the interests of the recipients; in many others, regulations or self-imposed codes of behavior helped to eliminate distortions and biases by encouraging the development of objective information or at least by clear identification of the biases. The precedents for protecting the recipients are well established; what is suggested here is not categorically new but a significant increase in emphasis.[33]

---

[6] The expression "target audience," common in television broadcast, is revealing in this regard. The recipients are viewed as "targets" for the transmitter's purposes, whatever they might be.

In addition to the content, the format of the information flow is also highly significant in regard to its impact on the recipients. A book, a pamphlet, or, more generally, printed textual messages require active cooperation, serious effort by the readers; they, in general, can and do sort out those they really *want* (not necessarily those they may really *need*). But when messages are carefully subjected to "audio-visual" treatment, ostensibly to facilitate communications, but in fact to break down the instinctive protection barriers of individuals against psychological invasion, many, and perhaps most, individuals become in many instances defenseless through saturation and fatigue.

The practical implementation of the information flow related aspects of our first premise should thus involve a dual thrust. First, the multiple processes now in existence aimed at improving the quality of information should be increasingly focused on the longer term, serious concerns of the society, and of the individuals. These include, but should not be limited to, those illustrated in Appendix[7]; typical are prohibition of harmful messages, warnings and disclosure requirements in regard to potential damage, identification of the message sources (and by implication, the inherent biases), professional restraints, and develop-ment and dissemination by public authorities, foundations, universities, and corporations of objective, factual information. Second, message formats and communication media should be encouraged to seek *active consent* of the potential recipients before, or as a condition to, entering into communication.[8] Based on currently observable trends, and in the absence of regulatory coercion, professional ethics of the communication industry offer some hope in this regard.

*

[7] Appendix—Note J.

[8] Appendix—Note K.

Further development of professionalism and of institutional restraints among those responsible for mass communications should be fostered as the principal means whereby conflicts inherent in the quality control of information flow can be attenuated.

There âre many instances where the exercise of unlimited individual freedom impinges on the interests of others, or quite possibly on those of rather large population segments. When the conflict is obvious and torts can be adjudicated according to established precedent, civil and criminal proceedings are called upon for redress. When specialized background is required in litigation and criminal proceedings, the support of expert witnesses is available to all sides.

When the possible actions (otherwise part of legitimate pursuits) of certain classes of individuals potentially affect the interests of the community as a whole, *institutional restraints* have evolved in order to preclude collective damage before its actual occurrence. Such restraints in principle limit the freedom of actions of the members of the institutional group but they are freely accepted in exchange for explicit and often legally sanctioned privileges. Thus, members of the medical profession endowed with the power of administering potentially lethal drugs and of exercising potentially catastrophic surgical skills are in fact responsible for immense suffering or its avoidance, dependent on the wise application of their skills and knowledge. Not unexpectedly, they are subject to an impressive array of ethical and professional restraints; they are also responsible for their own recruitment, development, and for the setting of standards of professional behavior. The need for professionalization and for institutional restraints in this case is well substantiated and was, in fact, accepted well before Hippocrates. [9] Lawyers, engineers, architects, professors, and ministers are being subjected to the same

---

[9]   The first documented evidence of professional regulation of surgeons dates back to 1700 B.C.; the Hammurabi Code even sets the compensation scale for the victims of malpractice and the penalties for the inept or unlucky practitioner.[34]

type of professional restraints; as other avocations reach professional status (e.g., nurses) they gradually establish their own ethical groundrules.[10]

If we accept the proposition that the information flow is an essential tool for social bonding and that its pertinency, reliability, and timeliness are essential elements in stability of future societies (just as its availability was essential at the initial evolutionary steps of democracy), and if we also accept the converse, i.e., that large volumes of information flow, containing a high proportion of inaccurate or deliberately misleading messages and presented in highly pervasive formats contribute to societal instability, it follows that those having responsibility for, and influence over, mass communication should be increasingly subjected to ethical and institutional restraints.

In a free society the freedom of individual expression should not be abridged, except in flagrant cases of obvious damage to the potential recipients. But mass diffusion of information should be subjected to peer scrutiny for content, quality, and format as a matter of course and limitations should be imposed, by voluntary professional rules, whenever the broadly interpreted interests of society as a whole are being put at risk.

There are ample precedents to support restrictive qualifications for entrance into any profession as well as self-imposed restraints specifically applicable to mass communications. Thus, one cannot exercise the profession of lawyer without having impeccable educational credentials and unimpeachable moral character; in addition, bar examinations essentially subject prospective practitioners to the state-sanctioned veto of their professional elders. There seems to be no serious reason why publishers and editors of all mass media (print, radio, and television broadcast) could not be subjected to the same procedures. As to content, there are many instances of self-imposed

---

[10] Many subprofessional occupations have licensing or "Board of Examiners" type standards which, at a lesser level of skill complexity, perform the same function.

restraints, e.g., broadcasting codes, television "seal of approval"; movie program ratings; many publications reject certain type or form of advertisements; there also seems to be a rather complete consensus to eliminate obscenity, vulgarity, and news items which, in the judgment of the editors is "not fit to print." Without any modification of substance of the existing structure of the mass communication industry, one could envision the increasing prevalence of professionally sanctioned codes of self-restraint where the judgment criteria would be biased in the interests of long-term societal cohesion and stability. Perhaps just a clearer understanding by the profession as to where these interests are currently being endangered would accomplish much toward the end in view.

*

Expertise in broader areas relevant to long-term societal problems should be made available to public decision makers, including the judiciary.

The need for exploring solutions to society's long-term problems, well before they reach the disruptive crisis stage, was mentioned or implied on several occasions. Many of these problems are incompletely defined and, to the extent that they are, solutions are elusive and controversial, to say the least—even before new stresses due to implementation of new decisions begin to appear. No single group or individual, acting on genius, inspiration or particularly competent assessment of past experience, can presume to hold the key to solutions or even to define noncontroversial compromises. What we can hope for, though, is that the several groups which do possess the ability to attack problems of this type and those which do (or claim to) have the responsibility for guiding society, would eventually recognize the need to fit their actions into a more coherent, long-term framework, modified in the detail when the circumstances warrant, but nonetheless always present as a reference for our value judgments and decisions.

Many, if not most, of the longer term social problems are associated with conflicts. The political process, which is at the core of democratic societies, has responded to the need for conflict resolution: elected representatives attempting to persuade each other about the merits of their respective viewpoints; decisions are carried by some predetermined arithmetic rule, simple majority, two-third majority, plurality, etc. As the problems become more complex and call on increasing degree of specialization, all legislative bodies, but more particularly the U.S. Congress, have attracted professional staffs composed of generalists as well as specialists in disciplines not exclusively connected with the legal profession.[11] But, as pointed out earlier, conflicts are not really solved at the law-making stage; many stresses appear far downstream in the implementation process, hence the proliferation of judicial decisions and sometimes even the judiciary taking over specific managment functions.[12]

Even in clear-cut cases, where the law and the facts of a controversy would logically suggest predictable outcomes, the judicial process long ago recognized the need for inter- pretation; even in matters brought before the august Supreme Court, the rulings are supported with what is humbly and honestly called opinions. Regardless of their legal expertise or devotion to Constitutional principles, the Supreme Court Justices (and those participating at the lower levels of Federal and State courts) are individuals steeped in their own social background and immersed in the intellectual and philosophical climate of their times.[13] As their decisions shape the implementation aspects of many legislative actions, as the issues they address become more complex and more involved in many unexplored social and

---

[11] In addition to the staffs assigned to individual Senators and Congressmen, the staffs of the General Accounting Office and of the Congressional Research Service enjoy well-deserved reputations.

[12] There have been many instances where the courts have in fact taken over the management of school systems in desegregation cases (Louisville, 1977-1978; Richmond, 1971-72); in at least one instance a Federal District Court has even determined the day-to-day fishing rights allocations and the opening dates of the fishing season (Washington, 1977).

[13] Cf. Chapter II, footnote 3.

technological ramifications, the collective wisdom of the courts would in all likelihood greatly benefit from the multidisciplinary expertise that is increasingly (albeit insufficiently) available to the Executive and the Legislative branches.[35]

For matters mostly involving problems associated with applications of new technology, the Office of Technology Assessment (OTA) was established in 1972 as a staff organization to support Congress. Its charter consists in pursuing scientific and technological policy analysis in response to inquiries by legislative committees, together with a modicum of latitude for independent initiatives. OTA's areas of concern range from radio frequency spectrum allocation through national information banks, applied genetics, population growth, all the way to Soviet energy policy alternatives. The OTA staff systematically avoids policy formulation (as contrasted to analysis) and thus has acquired the reputation of a highly respected center for in-depth professional expertise and for carefully nurtured objectivity.

It may be advisable to strengthen and broaden the staff support of the judiciary in areas beyond legal expertise, either by providing their own specialized staffs or by calling on the staffs of other Governmental branches. The operation of OTA or of the Congressional Research Service could be duplicated, or at least expanded, to serve the needs of the U.S. Supreme Court, for instance, and eventually also encompass those of the lower courts.

*** *

Governments at all levels should provide initiative and stimulus for new undertakings with desirable long-term societal objectives when the magnitudes, the payoff periods, or the financial risks place them beyond the normal purview of private enterprise.[14]

---

[14] Again, the U.S. Federal Government is used as an illustration. Other national, provincial, and state governments should be subjected to the same type of scrutiny from this viewpoint.

Since its very inception, as mandated by the Constitution and as driven by the perception of public need, the Government accepted (and probably welcomed) its increasing role in the economic and social aspects of the national community. As some of these governmental activities become more widespread, reasonably successful and accepted as part of the social fabric of the nation, they develop a momentum of growth of their very own. Those in charge enjoy the appearance, and often the reality, of power; those deriving benefits, directly or indirectly through the first and the succeeding generations; those who mold their political attitudes in the hope of favors and patronage—they all contribute to such growth. With society becoming more complex,[15] opportunities arise in large numbers, stimulated by conceptual or technical innovations as well as by the evolving public attitudes, perceptions, and expectations.[16] In the more mature stages, the increased role of governments is sublimated into a philosophical belief to the effect that public, rather than private, enterprise is the proper answer to many, and possibly most, of society's relentlessly proliferating problems.

Let us leave to a later section the public vs. private initiative controversy. The point to keep in mind here is that there are a number of important newly emerging societal needs, which, because of their nature, are not yet fully perceived as such; for which, in view of their nature or magnitude, governments (in concert with private industry or individuals) should take the initiative; and which, if left

---

[15] The role of information flow was repeatedly pointed out earlier. Public authorities, with the help of those in position to benefit from new governmental undertakings, are often remarkably subtle and effective in pointing out the unmitigated blessings accruing to the people from their insight, initiative, and competent stewardship. Perhaps one of the most revealing recent instances was the effort by the NASA in the late 1960's to conduct a systematic in-house exploratory effort, followed by intense nationwide information campaign, to identify and explain the benefits of space-related activities to the "man-in-the-street." (The Equal Rights Amendment having in those days not yet received the publicity of later years, women either did not count, or were endowed of superior degree of understanding, or perhaps simply were not supposed to be in the street; but at any rate, the campaign was not explicitly aimed at them.)

unattended, are apt to exacerbate the previously described stresses to the point of possible irreversibility. The broad problem areas are discussed under the headings of *controlled autarky, stable economic growth,* and *societal resilience.* This is not to suggest that these encompass all the areas where public initiative might be desirable; most certainly no case can be made for originality or uniqueness of such grouping. But they may serve as a useful set of illustrative examples.

Controlled Autarky [17]

The concept of controlled, as contrasted to maximum level of practical, autarky embodies the Government's responsibility to determine (through the normal political processes) the desirable mid- and long-term economic characteristics of the nation, as embedded in the matrix of international protagonists, allies and competitors alike. In concert with the private sector, the coordinated economic foreign policies and domestic capital investment policies of the nation should be aimed at supplying the raw materials, products, and services in support of the projected needs.[18] The degree of autarky in any given sector is *controlled* in the sense that it is not predetermined by theoretical principles or by mostly political considerations, but as part of a three-dimensional projection of the future: social, political, and economic. The latter includes, as a matter of course, the technology projections. The degree of involvement of private enterprise is not at issue here; just as in matters of national security, it is the Government's responsibility (using any expertise it can muster, including business) to determine future goals and broad requirements; at the implementation stage, private enterprise should compete for whatever share it can profitably underwrite.

16   Appendix—Note L.

17   Economic self-sufficiency as a national policy (Webster).

18   Appendix—Note M.

Stable Economic Growth

Prosperity and economic security as perceived by a large majority of citizens is, in general, accepted as a necessary condition to political stability. One might add, in the light of recent experience, that groups which are not at a given period enjoying that perception should at least be genuinely persuaded that society offers them a fair opportunity for advancement through their own efforts.[19] Even as the net population growth rate decreases in all the industrialized nations, the total disposable wealth must continue to grow at a rate consistent with the expectations of per capita growth of disposable wealth in real terms (adjusted for inflation). Any democratically elected government which, in absence of real international crisis, fails to satisfy this expectation faces swift political oblivion.

Growth, as envisioned here, is not necessarily associated with proportional increase in the use of energy or nonrenewable resources. Well to the contrary, it may be perfectly consonant with relative *decrease* in energy consumption per unit GNP due to conservation measures or in relative *decrease* in resource requirements per unit GNP due to recycling or renewal. Growth can be experienced by increasing the skilled labor content of the national output; by shifting consumption toward the highly skilled service industries such as art, leisure, and education; reduction of unnecessary obsolescence in consumer capital goods; by capital investments aimed at modernizing or extending productive facilities and infrastructure; by investment of labor and capital in the primary resource producing industries such as mining, agriculture, sylviculture, and mariculture; just to cite a few instances. In the present context, investments related to resource development and infrastructure improvement are emphasized. This is probably due to our current level of understanding since projects requiring relatively huge, visible facilities have been traditionally favored by politicians and planners alike. There is at least a possibility that the other elements of

---

[19] Relative prosperity obtained by means of transfer payments does not lead to stability, except as a near-term palliative.

"growth," requiring changes in individual attitudes and preferences, would contribute even more to long-term stability.[36]

Essential to any economic stabilization program along these lines is the availability of investment and working capital at the scale required by nationwide undertakings of this type. All the financing mechanisms cannot be examined here, but are competently discussed in the literature.[37] Roughly speaking, only three sources of funds are available: (1) taxes, excise, and custom duties levied on ongoing economic activity; (2) disposal or alienation of publicly owned assets; and (3) domestic and foreign borrowing.[20] These could be used in combination for the purposes sought here, provided that they intrinsically contribute to the stability objective, rather than further accentuate the causes of instability. If taxation is used, it should encourage production and investment rather than further stimulating resource-intensive consumption. Disposal or alienation of national assets should be tied preferably to clauses leading to substantial increases in capital value as a result of the disposal; the history of land grant financing may serve as useful precedent. Loans and equity capital could be attracted at relatively moderate cost and risk to the Government by first ascertaining the long-term economic soundness of the stabilization projects and, second, by resorting to Government profitability/risk insurance rather than to direct subsidies.

In addition to appropriate goal-setting and financing, the overall objectives of any large-scale stabilization program must rest on solid technical foundations. Well before entering long-term irreversible commitments, technology development and impact assessment type programs must be carried out as independently as possible from the biases inherent in the preservation of the status quo. Universities, nonprofit institutions, and (why not?) Government technical agencies should play a leading role in planning and conducting such programs. Recent examples in water management, environmental protection, energy sources, and

[20] Deficit financing is not considered adequate in these inflationary times.

nonchemical insect control offer valuable pointers in this direction.[38]

Finally, when assessing the benefits suggested here, one should keep in mind two further considerations. Technology, management know-how, equipment, installations, and even end-products of stabilization programs may in themselves become valuable export commodities. The needs perceived in any one country or region may find important applications elsewhere.[21] In addition, as we will have the opportunity to observe later in this study, stabilization programs also interact beneficially with other objectives, such as manpower development and social resiliency.

Societal Resiliency

The vulnerability of complex, interdependent societies, characteristic of the advanced industrial nations, has been pointed out earlier. It includes vulnerability to military action, which in an age of long-range nuclear weapons assumes especially awesome dimensions; it covers exposure to lesser levels of violence such as terrorism and rioting which use perhaps much smaller amounts of destructive energy, but a much higher degree of information and malice and, for this reason, are apt to be even more dangerous. Except for a few unfortunate cities in World War II, the current generation of metropolises, in particular in the U.S., have been planned and constructed without serious regard to military and civilian unrest related risks. The protection of civilian life, the preservation of raw materials and supplies, the maintenance of law and order, communications, and the recovery potential of industry have received only the most perfunctory attention.

One area of societal investment, which at this point would not compete with the normal economic life of nations, would provide the organization, capital investment, and operating funds required to implement long-term societal resilience programs, including but not limited to civil defense in a warlike situation. It would presumably

[21] Appendix—Note N.

contribute to its primary objective; i.e., to improve the protection and recovery capability of the population and its economic activity following hazards associated with wars, violence, and disasters; but it would also provide an economic "inertia wheel" for surplus skills and labor as well as for commodities, semifinished products, and capital goods without competing with the normal course of the national economy. It could thus, by the proper modulation of the rate of investment, serve as an important element in stabilization.[22]

* * *

Public policies and private initiatives should conjointly aim at providing economically justified roles for marginal producers. The reward/penalty structure of democratic societies should offer strong incentives for active partici-pation in the economic process.

The adverse social consequences of this growing "marginality" trend in the labor force have been observed earlier.[23] Democratically elected governments character-istically tend to apply the near-term remedies rather than attacking the causes: Welfare, unemployment, subsidies to sectors impacted by competition are among the favorite approaches even though repeatedly proven expensive and ineffective. In special cases, when basic skills and motiva-tion are at hand, retraining programs have experienced some limited success, especially when under the operating control of private industrial organizations. In the majority of cases, however, training programs pursued by the Government are little more than make-work projects or thinly disguised welfare subsidies. The proportion of marginal workers actually and permanently returned to the productive main-stream remains disappointingly small.

---

22 Appendix—Note P.

23 Chapter II.

Few are the approaches which offer radical departure from the past with any real hope of success at expenditure levels affordable for the long haul.   But one may reasonably surmise that once the policy objective of providing an economically justifiable role for marginal producers is accepted, the combination of some of the following suggestions may lead to significant improvement:

● Use the large-scale capital investment projects undertaken for the purposes of economic stabilization as training ground for the unskilled, or as retraining of other marginal workers in marketable skills;

● Exploit the possible synergies in connection with other desirable undertakings such as urban renewal, protection and improvement of natural environment, and wildlife habitats; [24]

● Shift increased proportion of available subsidies, grants, research projects, and loan funds to vocational training;

● Use training programs currently being pursued by the Armed Services and the National Guard for skill development or retraining. Not only would this result in the direct upgrading of the corresponding labor segment, but it would also offer the opportunity for motivating the most promising trainees for longer term military-related assignments. In addition, a larger proportion of the work force may become available for tasks requiring specialized skills in case of national emergencies;

● To the extent that transfer payments are available for alleviation of marginality-related hardships, channel these funds through industrial and business organizations. Let private industry use these resources to select, train, and eventually employ those it really needs;

---

[24] Projects of this type were quite successful during the 1929-1936 depression.

● Exploit the synergies between marginal groups of workers. In many areas, for instance, teachers are in surplus; instead of trying to foist this potentially valuable resource on school systems that are unwilling or unable to make constructive use of it, one may think of retraining first the surplus teachers, social workers, etc., and then motivate them to start providing training services to other marginal groups, under Government or industry-sponsored programs;

● Instead of emphasizing the development of highly auto-mated machinery in the name of production efficiency, both Government and industry should start initiatives for the development of *interface machinery.* This term is applied to equipment which, when coupled to rela-tively unskilled or marginal personnel, produces sophisticated goods and services, intrinsically valuable or not otherwise available.[25]

All these, singly or in combination or perhaps augmented by other ideas attempting to accomplish the same purpose, may indeed prove beneficial. But it should be obvious to anyone who has observed the evolution of society since the early 1960's that success is not likely to follow unless the basic motivational elements among the marginal workers are strengthened or reversed in the direction strongly favoring productive and responsible participation. When public policies and attitudes condone, mitigate, and (one is almost tempted to say) encourage alienation; when they in effect finance the voluntary "opting out"; almost create the excuses or rationalizations why one should not work and in effect eliminate the dire consequences of not working...let us face it, a significant proportion of actual and potential "marginals" will attempt to take permanent advantage of the system.

Harsh as this may sound, public sympathy and support should be strictly reserved for individuals who are physically or mentally unable to provide responsibly for their own support. Any able-bodied person (except those with depen-

[25] Appendix—Note R.

dent children of preschool age) should work for his or her rightful share of necessities such as food, clothing, and shelter, preferably paid directly to the suppliers. If the "dignity" of the recipients suffers because of the somewhat inquisitive scrutiny, so be it; this would further add to the positive motivation for finding and keeping a job. There should be a strong and increasing social stigma attached to receiving welfare benefits without well-authenticated cause.

On the other hand, the positive motivations should also be explored and strengthened. What immediately comes to mind are the taxes and social security paid by individuals who just have left the unemployed or welfare category and are in the process of developing the attitudes required to become permanent members of the productive population. Why not offer them a tax moratorium of, say, five years so as to put a real premium on working, rather than making it a barely paying proposition? Many other ideas should be explored in this direction.

*

The systematic development of knowledge aimed at problems of societal cohesion and stability in the industrial and post-industrial stages should be recognized as an urgent national need and supported accordingly.

The conscious adaptation implied in the hypothesis formulated earlier[26] requires adequate and timely perception of the changes in the environment, followed by decision and implementation processes that are rational in terms of a priority-ordered set of societal goals. In relatively simple societies this can happen (and indeed must have happened in the past) when the wisdom and experience of the leaders are sufficiently relevant to cope with relatively slow environmental changes. Unfortunately, as I have argued earlier, the complexity of the problems facing a modern industrialized society and the rapid pace of major changes in our environment often defy traditional wisdom and render at

[26] Part 1, Chapter III.

least some of our collective experience irrelevant, obsolete, or even downright misleading.

Thus, one of the basic functional problems faced by typically complex societies is related to the prompt and objective perception of the environmental trends so as to *predict* the changes rather than to observe and suffer the consequences.   Coupled to the prediction, competent decision models should be at hand in order to assess the alternate available courses of action. This relatively simple requirement may sound naive or utopian, but let us not forget that a large number of public and private decisions are made each day on the implied premise that, in the aggregate, they will lead to beneficial consequences. So an attempt to improve policy decisions, at least those which are fraught with potentially momentous consequences, is perhaps not entirely irrational.

Western societies have faced this type of problem in the past, more than once. The mid-seventeenth century brought to Europe the first impacts of "hard sciences" (mathematics, physics, astronomy), and those of reliable long-distance navigation. The proto-industrial West, characteristically, responded to the intellectual challenge by evolving their version of modern think-tanks: the Royal Society for the promotion of Sciences and useful Arts, established in England in 1667 was soon emulated in France, Russia, Sweden, and (in the following century) by almost all modern nations, including the colonies of North America.[27] Assembled initially by shared curiosity, soon these learned societies were called upon to solve problems of urgent interest to their respective governments and, more generally, to serve as scientific and technical advisors to policy formulation and as steering groups or even managers for national undertakings of critical priority and import.[28]

[27] The American Philosophical Society, informal equivalent of the Royal Society, was founded in Philadelphia in 1744.

[28] In England and France the problem of determining longitudes at sea was one of the significant examples when scientists ("philosophers") and technicians ("mechanicks") cooperated for the benefit of the early version of the military-industrial complex. Similar instances could be cited in the

Closer to our time, practically all Western nations vie with each other in the chartering of a remarkably diverse multitude of problem solving organizations.   In the U.S. the Government has a few mostly scientific and technical agencies such as the Bureau of Standards, NASA (formerly NACA); but almost all Executive Departments such as DOE (having absorbed the AEC), Human Resources and Education (formerly HEW), Agriculture, Labor, have research and development related activities aimed at specialized problem solving.   Congress has established the Office of Technology Assessment; universities maintain numerous institutes and centers for political science, "peace research,"[29] government science, in addition to their departments of social sciences; many private not-for-profit organizations, such the The Brookings Institution, the Hudson Institute, the Batelle Institute, and the Rand Corporation, intensively engage in similar activities.   Private foundations, such as Ford, Carnegie, Mellon, were in the past, and still are, at the forefront in sponsoring studies of broad social relevance, especially where the specific approaches do not fully coincide with the preconceptions of the Government or with those of the established academic disciplines.

There is thus no shortage of professional talent in the field of political and social sciences, economics, technology and in many other, possibly relevant, fields of specialization. What perhaps could be suggested is that the many individuals and agencies, currently pursuing research in their respective areas of interest, be brought together (preferably without formal Government sponsorships) in a series of workshops and be charged with the task of developing an institutional framework to attack the broad problems of preserving stability and cohesion in industrial and post-industrial societies, with the overriding condition of

28   (Footnote continued)
   American Civil War, in the Franco-British effort to attack the antisubmarine warfare problem in World War II, in the U.S. "Manhattan Project." In the Soviet Union the Academy of Sciences actually supervises the nation's efforts in nuclear fusion energy, in molecular biology, and other important areas.

29   e.g., The Hoover Institution for War, Revolution, and Peace at Stanford University, one of the earliest prototypes.

protecting individual freedom and human rights. Representatives of other Western nations should be welcomed both as observers and participants.

* * *

Antagonisms among Government, business, labor, and professions should be attenuated by deliberately designed policies and rules of governance applicable to all.

Many of the foregoing sections suggest, or at least imply, new initiatives to be taken and new responsibilities to be assumed by the Government. In view of the many past failures at both the policy-making and the implementation level, (some of them vigorously decried in earlier chapters) any hint about further increasing the role of Government may sound illogical and inconsistent at best and would probably be dismissed as doctrinally inspired nonsense. Clearly, a few words of justification are in order.

In all types of societies, but more particularly under the pluralistic democracy characteristic of the U.S., political and economic power is enhanced by joint efforts. Individuals with kindred interests will increase their influence by concerted action until checked by countervailing interests. Such an equilibrium among societal groups owes its stability to a judicious admixture of cooperation and competition;[30] the "adversary" relationship between Government and business often observed in the U.S. (and to a lesser extent, the interactions with labor and the professionals) are nothing more than specific manifestations of this general rule.

The extreme views held and publicized by some of the Administration agencies and officials, or those presented to often sympathetic Congressional hearings by the self-anointed protectors of the consumers about the objectives,

---

[30] Unconditional and permanent cooperation would cause the neighboring groups to coalesce in fact and thus jointly increase the competitive pressure on those external to this new "joint-group" entity; unrestrained competition would soon result in the weakening of the competitors and their subsequent subjugation to, or absorption by, neighboring groups.

ethics and practices of business are, to put it mildly, unrepresentative of the general truth.   Business, big or small, evolved over centuries and (in the modern Western form) over several decades; it certainly harkens back to periods when private property and profit were protected with almost religious fervor, when ecology was (at best) an arcane field of study relegated to obscure academic departments and when "social consciousness" was mostly manifested by the workhouse and parish charity.   To describe typical business managers as a class of greedy, narrow-minded tyrants, engaged as a matter of sustained interest in "ritualized thievery"[31] is just (or perhaps, almost) as unfair as to form a general opinion about the hierarchy of the Roman Catholic Church based on the behavior of the Borgias.   It would be an exaggeration, of course, in the opposite direction to support the view that senior business executives are, as a group, more generous, altruistic, humane, or insightful than members of other groups at roughly the same socio-economic level.

Balance, tempered with a measure of charitable understanding, should be our motto as well when assessing the performance of governments.   As individuals, legislators, members of the judiciary, and officials of the Administration are probably quite representative of the talents and virtues, as well as of flaws and failings, of their peers engaged in business or professional pursuits. To suggest that civil servants are, as a group, mindless and incompetent morons parasitically feeding on the substance of the hard-working population[39] is again just as unfair as to support the view that officials, by the sole virtue of their position, are endowed by the superior wisdom and foresight not usually vouchsafed on lesser mortals. In truth, modern governments attempt to solve, perhaps with a trifle of unwarranted zeal, complex problems which are often incompletely (and probably incorrectly) defined.   They must rely on the dubious help of mostly deficient theoretical and empirical knowledge, often in areas where deliberate experimentation is not exactly encouraged. Failures, or even temporary setbacks, are harshly penalized by the political process—

[31] Steinbeck.

more so perhaps than in the much-advertised "jungle" of the business world.

If any further argument be found wanting in support of the view that neither government nor business is categorically competent to "solve" most, or at least certain important classes, of societal problems, let us pause to observe that policy-level Government officials, members of Congress, and of state legislatures often are transfuges from the business or professional worlds; conversely, it is not exactly unusual to see senior Government officials reincarnated in the plush precincts of corporate board rooms following retirement from their public career. Business expertise at the senior government level is thus fully available (although not easily brought to bear on day-to-day implementation problems) and the intricate workings of government are often familiar to those setting the pace in the business world. (It is not obvious, unfortunately, that such familiarity breeds mutual respect and beneficial intercourse.)

The view is taken here that no single segment of society (government, business, labor, or professions) has the intrinsic competency or the exclusive responsibility to unilaterally determine and implement policy changes in response to the emerging problems of cohesion and stability. Government should determine policy and establish long-range goals, by using all the contributions from business, labor, and professions; it should direct, sponsor, and coordinate the analytical and fact-finding efforts aimed at evaluation and implementation. It should submit its findings to the arbitration of the political process (in fact it has little choice but to do so) with the added burden of ensuring that the arguments of all sides bearing on important issues are given full and objective hearing. Foundations, with their access to considerable financial and intellectual resources and with the close cooperation of the "not-for-profit" institutions, can play an important role by ensuring through their own efforts that Government sponsorship of the policy formulation process does not lead to political bias and distortion. Special interest groups of all persuasions, including those who take strong views (pro and con) on the problems of growth, environment, energy, wildlife, abortion, fluoridation, SAT's, SALT, and other causes innumerable, should be offered respectful hearing, without any of them being given the opportunity to impugn the motivations or the competency of their opponents. Business, as represented by individuals and "corporate citizens," should also bring its resources of motivation and ingenuity to bear on the proposed implementation phases. The experience in innovation and the relatively rapid information gathering capability of business are potentially most valuable adjuncts to Government efforts.

When stated in this (admittedly idealistic) simplicity, few people would raise objections of substance to the proposition that Government and business should cooperate for common benefit. Such cooperation exists in fact in all Western nations;[32] its scope ranges from full or partial ownership through joint Government-private ownership of corporations, subsidies, loan guaranties, tax preferences, research services, all the way to full private ownership favored only by export financing and other manifestations of Government approval. When one examines the facts, it becomes quite apparent that close (and in the whole, beneficial) cooperation between Government and business is the norm even in the U.S., the many small irritations associated with regulation and taxation notwithstanding. Without such cooperation our major Allies,[33] who do not seem to display the same fervent devotion to antitrust principles or the same moral abhorrence of politically motivated sales commissions, would even further enhance their competitive strength, when pitted against the U.S.

The definition of the respective spheres of responsibility and competency between business and Government should be clear, even though subject to minor controversies in the detail and to possible changes in the presence of new

[32] It exists, of course, in a somewhat less-than-spontaneous form in the "socialist" regimes where the government *de jure* owns and operates all major production and service sectors.

[33] Japan, West Germany, England, and France.

circumstances. It is healthy that both business and Government (and many other participants in this fascinating game) be alert in scrutinizing their respective performances. But hostile propaganda, deliberately orchestrated on both sides in order to curry favor with the electorate, should be eliminated. It does not help the nation[34] having the Government depict industry and business leaders as engaging in large-scale nefarious schemes to fleece the public, to destroy the environment, and to evade their taxes. Business should equally desist from portraying the Government as planning and implementing programs aimed at strangulation of business by taxation, over-regulation, and demagogical emphasis on ever-increasing benefits to nonproducers. Past instances of excesses and aberrations should be corrected (and it is surprising how often they *are* without real permanent damage) and the longer term cooperation should be accepted as a mutual goal, leading eventually to partnership truly beneficial to the national community.

***

Perhaps more than any other, this chapter justifies its being designated as part of an essay.[35]

Given the premise that the maintenance of societal cohesion and stability is an explicit goal of Western nations and that it should be pursued on par with the protection of individual citizen's rights, a few policy derivatives are presented. Some of them, such as the control of information flow, and the Government's role in providing initiative and stimulus for major new economic and socially oriented undertakings, are broad and raise rather basic issues; some others, such as the systematic development and application of pertinent background knowledge, are related to relatively noncontroversial matters of implementation.

---

34 Referring here to the U.S.

35 "Short literary composition, dealing with a single subject, usually from a personal viewpoint and without attempting completeness." (Webster)

The only common element among the topics discussed is that they are all logically connected, one should hope, to the central premise. They do not, by far, exhaust the subject; it is not even reasonable to claim that the most important aspects have been adequately identified and treated with the appropriate distribution of emphasis. But, taken as a group, they form a remarkable illustration of the conclusion that a number of policy changes can be envisioned *now* which would positively influence the future stability and cohesion of modern democratic societies, and that there is really nothing radically new or incompatible with Western political ideals in the implementation of such changes. We mostly argue (if the central premise is accepted) for a clear recognition and public articulation of the pertinent long-term goals and for enhancing and guiding many randomly oriented evolutionary changes *in the process of occurring now* into channels consistent with these goals.

The reader is entreated to consider whether or not any alternative premise might be more attractive (for any reason at all) than the one proposed as a heading of this chapter; to assess the desirability and practicality of the suggested policy derivatives; to offer, if so inclined, modifications, challenges, and (I express that hope with considerable diffidence) extensions in still other directions. Most importantly, I hope the reader accepts at this point my earnest assertion that unless each nation strengthens individually the cohesion of its own society, the prospects for long-sustained collective action are rather remote. But that (to quote Kipling) is another story, to be discussed in the next chapter.

CHAPTER XI
Commonwealth of "Human Democracies"

SECOND PREMISE: THE LONG-TERM INTERESTS SHARED IN COMMON BY THE WESTERN TYPE NATIONS DO (AND WILL) BY FAR OUTWEIGH THE CAUSES OF COMPETITION AND CONFLICT AMONGST THEM. THEIR GOVERNMENTS SHOULD PURSUE THE MAINTENANCE OF COHESION AND STABILITY OF THE GROUP OF THESE NATIONS AS AN ESSENTIAL ADJUNCT TO THEIR DOMESTIC ORIENTED NATIONAL GOALS.

The success of the Europe-based Western civilization between 1650 and 1900 is without precedent at the global scale and has been ascribed to many confluent causes.[1] That this civilization succeeded in subjugating or colonizing practically all of its competitors is proof of its intrinsic viability, especially when viewed against the background of almost permanent internal warfare. By the time the dynastic and religious wars subsided, quarrels between nation-states took their place with the associated ethnic, racial, and economic motivations. By the early 20th century, World War I (fought essentially for economic stakes, but clothed in the guise of dynastic competition as well) exhausted the resources, and broke the aggressive spirit, of the European colonial powers. The residual

---

[1] Climate, soil, and geography are (as always) the primary causes in emerging civilizations. With the introduction of food grains and domestic animals, with the invention of the sideboard plow, the chronic specter of starvation receded, at least in the most favored regions, as early as 1000-1200 A.D.; the many navigable rivers deeply penetrating the continent, the relatively sheltered ports and waters of the Mediterranean and the Baltic favored extensive navigation and added to the many land-based migrations.

stresses of that conflict (economic, social, ethnic) favored the rise of new "ideologies," advocating social systems fundamentally different from the Western democracies as they existed by the 1920's. The challenges from nazism, fascism, and communism to the West were the fundamental causes of World War II, with the superimposed conflict between the U.S. and Japan; the latter being a latecomer to imperial ambitions in the Western Pacific at the time when the U.S. and the European colonial powers were otherwise preoccupied.

The fallacies of the World War II "victory" of the Allies have been pointed out earlier.[2] The U.S., Western Europe, and Japan are now faced, together as a group, with the challenge to their political and military domination by the Soviet Union, and to their economic domination by the emerging Third World.

This chapter raises the issue whether or not the West can reasonably hope to meet this dual challenge while continuing unabated their internal competition; it also argues that concerted action based on supranational political organization is the more desirable long term approach, and that such organization is not only possible within our own time horizon, it is in point of fact imperative.

The systematic pursuit of explicitly stated common goals among the Western nations, evidenced by jointly defined actions, subduing local interests and residual distortions of history, may go a long way in the direction of alleviating the internal stresses of the participants. It should, therefore, be pursued as a prime national goal of all the nations involved, while recognizing that the problem of organizing concerted action among major nations with different and often hostile backgrounds will stay with us a long time, probably for a generation and more. Considerable difficulties are in the way, but no alternative solution appears to be promising. If the Western style of life is to survive, if our political ideals are to prevail, the task of achieving permanent supranational cooperation must be addressed.

---

[2] Chapter VIII.

The shared value judgments and threat perceptions, the emergence of social stresses due to the same basic causes suggest that the broadly interpreted interests of the Western nations overlap to a considerable extent. This may form the basis for long-term commitment to a common political system.

The causes of stresses discussed in previous chapters, such as the information environment, the social unrest, the problems of ecology, the economic instabilities, and the military threats, affect all Western nations to various extent; none can live in economic or ideological isolation. Even such traditionally neutral countries as Switzerland and Sweden are in fact strongly interdependent with, and tributaries of, the Western economies. Among Western countries the individual lifestyles converge rather rapidly. No matter what measure is used, such as GNP per capita, urbanization, productivity, working conditions, fringe benefits, or transfer payments—all are associated with problems which the governments must deal with on a day-to-day basis. Business and labor rather freely exchange information across international boundaries and it is to be expected that any substantial differences in productivity, wages, and fringe benefits between Western nations will be of relatively short duration.[3] On the other hand, differences in national leadership and discipline, ingrained cultural habits, capital accumulation, and organized labor activity may result in appreciably divergent growth rates, especially if unequal burdens of military preparedness are allowed to distort further the economic picture.

Even though details of the several political systems are quite different at the operating level, some basic principles

[3] Japan, by its current lifestyle and owing to its ability to absorb alien customs, should be considered as a fully Western-style nation, even though its ancient culture retains strong emotional ties to the East. Prior to World War II, failure of the West to create mutually beneficial economic relations with Japan cost some of the major damage resulting from World War II and its sequels. How deeply and permanently Japan will remain committed to its current course based on Western values, even when the combination of economic pressures and military threats (or opportunities) will have drastically curtailed its enviable record of sustained growth, is probably the crucial contingency facing the Western world.

are broadly shared: Governments are to be freely chosen by the governed; basic civil and human rights are not to be subordinated to the interests of the state except as defined by law and as enforced by due process; systems of government should not be imposed or propagated by means of subversion or conquest.

Western nations also share, although in different degrees, the perception of a common threat with three interacting prongs. The most obviously apparent is the increasing military might of the Soviets, continuously present, used whenever the necessity or the opportunity arises, and backed by carefully advertised strategic nuclear "equivalence" with the U.S., which makes deterrence of limited military initiatives against Western interests somewhat less than credible.

The ability of the Soviets to initiate, and to exploit political and economic perturbations, threatening access to vitally needed energy resources and raw materials, further exacerbates the intrinsically present North-South conflicts, especially when these are already complicated by ideological and religious convictions.

More fundamentally, the Western social philosophy is based on the individuals' personal motivation, *determined by their own value judgments,* as the key factor in economic and social success, with the concomitant distribution of rewards and privileges. The (individual or collective) ownership of working capital is part of this philosophy. All this, of course, is in sharp contrast with the views prevailing among "socialist" nations which basically assume that the state, by applying abstruse principles, is in position to determine who needs and receives what, and which make orthodox political beliefs the primary factor in determining individual rewards. As the economic growth of industrialized societies is being further constrained and as the effects of social stresses become more evident, the ideological challenge by "socialism" (as advocated by the Soviet Union and some of its satellites) is apt to introduce further elements of instability.

One may, without serious risk of controversy, identify the generally recognized areas where the Western national interests overlap as follows:

- Stable global economic systems including stability in the balance of trade, balance of payments, and currency exchange rates

- Access to markets and to raw materials on an economically rational basis (including oil in the near term and other critical raw materials in the longer term)

- Maintaining the continuing stable rate of improvement in the quality of life of Western nations in a relatively peaceful world by preferably confining the unavoidable conflicts to remote peripheral areas

- Elimination from the Western collective behavior those elements which may offer attractive options to the Soviets, if they persist in pursuing their hostile objectives, through a combination of diplomatic, economic, or military means

- Normalization of trade relations with the Peoples Republic of China, provided that the advantages of such trade outweigh the risks of further confrontation with the Soviet Union

- Sharing the burden associated with the integration of nations with large, fast-growing populations but with insufficient or inaccessible natural resources (the "Fourth World") into the world economic structure

Each Western nation faces the basic choice between two alternatives. They can enhance and exploit the favorable differentials so as to secure high growth and capital investment rates, but this "euphoric" stage is not likely to last very long; soon such nations would begin to suffer from competition and trade restrictions; soon the pressures would arise for increased military spending to protect the fruits of their national enterprise. The alternative is to cooperate with economically and politically kindred nations ("the West") and to share the burdens of such cooperation with the hope of enjoying its lasting benefits.

The basis for such coordinated political and economic action is clearly present now; it is proposed that a higher level political organization should replace the current system of ephemeral and shifting alliances. The Western nations have common economic interests and have to a large extent kindred goals and political traditions; they share a common or at least overlapping perception of the threats. The communications and the analytical capability are present to a sufficient extent to warrant the evolution of a political system deliberately designed to serve better their common purposes.

*

Political agglomeration is the historically correct solution, once the communications have evolved to the point of promoting common perception of mutual interests. Joint action toward the outside and attenuation of antagonisms inside are the principal advantages.

It is, of course, possible to dream that by some miraculous, as of yet unspecified, mechanism the industrialized nations will correctly assess their present situation and their future prospects; that, as sovereign nations acting in concert, they will evolve purposeful and effective policies and will then be afforded the leisure to implement joint action plans in a general atmosphere of mutual respect and understanding. Extrapolating past history and viewing the evolution of the geopolitical environment as it affects the Western nations *in toto*, it is probably fair to assess the chances of such spontaneous cooperation as being rather slender. It is more likely by far that the individual nations will persist, if for no other reason than to respond to their domestic short-term directed pressures, to compete and vie with each other over many secondary issues; that they will continue to muddle through crises as and when they arise; and that they will eventually face, individually and collectively, the prospects of upheavals or perhaps global-scale catastrophe. Even more likely is the eventuality that the cohesion and assertiveness of the industrial nations will gradually decay in close parallel with their declining productivity and quality-of-life, to the point that individual citizens might well stop being concerned about anything, let alone the future of their political institutions.

Historically speaking, the correct answer is political agglomeration.[4] Once the need is understood, the associated benefits and sacrifices evaluated and emotionally accepted, the governing segments, the opinion-makers, and the majority of populations will support both the idea and the reality of a supranational power preempting some of the important roles of theretofore sovereign governments. The degree or depth of agglomeration is less important in the initial phases than the principle;[5] once the populations get accustomed to the exercise of power in the name of the new entity, especially if such exercise is perceived as clearly beneficial, rather than unwarranted, arbitrary or pervasive, it becomes increasingly accepted and eventually supported.[6]

If the second premise, subject of this chapter, is accepted, then the national governments will... "pursue the maintenance of cohesion and stability of the group of Western nations as an essential adjunct to their domestic-oriented national goals." We are now to examine the proposition that the only practical long-term implementation policy derived from this premise should be aimed at definite, progressive

---

[4] This term is used to describe a political organization endowed with the legal right and the actual capability of overriding, in specifically circumscribed instances, the sovereignty of individual nations.

[5] It took 16 years following the Declaration of Independence for the U.S. to ratify the Constitution and the first ten Amendments (1791); then another 15 years or so before the apportionment of power between the Federal and State governments was broadly defined. It has been probed, criticized and amended ever since, but the general trend shows a continual increase in the executive powers vested in the Federal Government with the corresponding decrease in those retained by the individual states.

[6] The idea of a strong U.S. Federal Government was for a long time violently resisted by a large number of influential people otherwise strongly devoted to the idea of independence.

steps toward a supranational agglomeration of the Western nations. They should be *definite*, i.e., comprise specific commitment of the participants toward the common objective; they should be *progressive*, i.e., allow the time span necessary for individual political adjustments and for the *accretion* of the novel entity by the joining of new nation-members as motivated by the development of mutual interest and perceptions.

Alternatives do not appear as potentially rewarding. To remain in the *status quo* does little or nothing to alleviate the currently observable stresses burdening the West. If there is any proposition generally accepted by Western prognosticators (and gleefully reinforced by those of the Eastern bloc), it is that the West is doomed to gradual erosion and decline, even in the absence of violent conflicts. To proceed more rapidly than implied here in the direction of close union or federation-type political system is probably premature and unrealistic at this time.

Until and unless a more expressive and appealing designation can be found, the expression "Commonwealth of Human Democracies" is submitted on a trial basis. We should eschew the exclusive "Western" orientation, since the earnest hope (and the only chance) for enduring success is based on the appeal to *all* nations, specifically including Japan and others as their political evolution and alignments permit. "Commonwealth" is to emphasize the community of ideals and interests, as contradistinguished from liens imposed by purely political circumstances; the word "Human" intends to convey the essentially shared belief in the primacy of human rights and dignity in formulating the domestic social institutions of the member-nations. We shall, in what follows, consistently refer to the proposed supranational authority as the Commonwealth, being understood that the name should be considered as a temporary expedient.

Let us first briefly recapitulate the arguments in favor of a Commonwealth encompassing all Western nations:

1. From the viewpoint of any prospective member-nation, the internal competition among the participants will be restrained by setting standards for wages, social overhead charges, migration groundrules; the residual competitive factors will mostly include intrinsic elements of high productivity (raw materials, skills, capital facilities); this will make the Commonwealth, as a group, more competitive against those at the outside. As part of the Commonwealth, the Western nations would no longer use their individual interests as the basic criterion for interaction with external groups, weakening or destroying thereby their common leverage.   This holds against the Soviet Union in the future competition for raw materials and for opportunities of technology transfer; it is even more critical with respect to the Third World. In this latter perspective, the competition for markets and raw materials should be among the most salient aspects of international trade to be subject to the common authority.

2. With full Western solidarity, valuable policies, which serve the advantage of each individual nation, would be supported in international negotiations by a single voice, thus lessening the bargaining power of opponents. This is applicable to the strategic arms limitation treaties, to the definition of nuclear-free zones, to negotiations affecting mutual balanced force reductions, to the prohibition of nuclear weapon proliferation, to the definition of "the law of the sea," and to a number of other internationally oriented enterprises and conventions.[7]

3. From the viewpoint of Western nations (U.S. excepted) the essential advantage of the political agglomeration implied by the Commonwealth is to achieve a situation where they, as a group, can exert legal and binding influence (encouragement or restraint, as the case may be) on the economic and foreign policy decisions of the United States. To the extent that such influence exists

---

[7] International use of communication, navigation, and meteorological satellites and allocation of radio-frequency bands are examples of current interest.

now, it can be only exerted after the fact, creating thereby delays, sources of conflicts, and irritations. The net effect is confusion and ineffectiveness, but quite often the Allies have to choose between publicly dissociating themselves from U.S. policy or suffering the consequences of policy initiatives which they perceive as ill-advised or opposite to their national interest. In effect, owing to the economic and military preponderance of the U.S. in the Western world *and* to the self-assigned role of the U.S. in pursuing its global responsibilities, the Allies are often drawn into conflicts without adequate representation or consent.

4. From the viewpoint of the U.S., supranational Commonwealth authority would in effect constrain to some extent the options and initiatives which would otherwise be available to its foreign policy makers as officials of a sovereign nation. On the other hand, such constraint would be more than compensated by the legally binding support of the Western Allies, once a collective decision is reached by the properly constituted Commonwealth authority.[8]

5. Far beyond these reasons, which at best would just alleviate the day-to-day conduct of joint Western foreign affairs, there is a fundamental decision facing the U.S. in particular. All our long-range goals are focused upon the preservation of our socio-political system and our "way of life," preferably within a peaceful, stable world. We can approach this task with the attitude of *dominant leadership* with all the necessary wisdom (which we may have to acquire) and all the required domestic discipline and sacrifices (which we will have to accept). Alternatively, we can perceive that our goals are inextricably embedded in the preservation of the Western world, organized as a viable, dynamic, and cohesive entity, capable of facing the

---

8 The U.S. public should not delude itself. In fact, even now the major U.S. policy decisions are constrained by the needs and attitudes of the Allies. The delays possibly incurred by the collective decision mechanism may well be compensated for by the long-term viability of the results.

present and future challenges with well-founded confidence of success.   The Western nations, acting in concert to pursue their common objectives, would jointly bear the burdens for their economic stability and military security; they would correspondingly share and enjoy the benefits or, paraphrasing consecrated words, "...secure the Blessings of Liberty to themselves and to their Posterity..."

If we choose the second approach, each nation, including the U.S., will sacrifice a significant share of its sovereignty by subordinating to that extent its national interests to those of the community.   At that price, the major contingency that Western Europe, Japan, or other important Western assets would in fact be neutralized or subjugated by hostile power blocs would be averted.   This single consideration is thought to be so important that it fully justifies the commitment of the U.S. to the Commonwealth principle. Should Western Europe, through violent conquest, intimidation, or through apathy, following a period of economic decay and ideological confusion, fall into the Soviet orbit, the future of the U.S. would be very bleak indeed. The same type of ominous long-term threat lurks in the Far East, should Japan, for any reason, under duress or modified perception of its political future, cast its lot with the PRC.[9]

*

---

[9]   The currently popular "China Card" should not be looked upon as a beneficial long-term move for the West. The PRC may win friends in the West by offering markets for industrial products, by encouraging capital investment, and by publicizing the (relative) relinquishment of some of the more restrictive features of their society. In the long run, though, if a major war with the Soviet Union is averted, Red China will definitely not evolve into a benevolent pacifist giant, satisfied with its role as a "village community" at the global scale, catering to the needs of the "cities" of the Western world. It will eventually grow into the posture of a nuclear-capable superpower, perhaps less expansive and ideologically assertive than the Soviet Union, but definitely an aggressive competitor for the ever-precious raw materials and perhaps even energy resources. For Japan, the possibility of overcoming its historical and racial antagonism against the Chinese and of supplying the technological and managerial leadership to accelerate the emancipation process is a very real one, no matter how misguided or unpalatable it may appear from the Western viewpoint.[40]

The evolution of supranational political authority is a slow process at best—it may take another decade or a generation before it can be relied upon as a positive force in world affairs.

When contemplating the prospects of a successful Commonwealth, we might as well realize right at the onset that the task of eliminating centuries-old traditions of mutual hostility, and the problem of overcoming the ever-present incentives among the ruling elites to preserve the status quo, are by no means easy ones.   There may be interest in briefly reviewing all the negative factors which may bear on the evolution of concerted political action, leading eventually to supranational organization.

As a general observation, nations will change their societal habits slowly, gradually, and then only under the effects of clearly perceptible pressures.   Abrupt changes in the form of governments or in the relationships affecting sovereignty are even more difficult to bring about; in general they would only result from violent confrontations rather than from mature and competent weighing of the alternatives.   Whatever changes the Commonwealth concept may require must be gradual and introduced on an ad-hoc basis of perceived needs, without perturbation of the form or substance of societal habits engrained in the individual nations.

One of the basic difficulties in establishing permanent multilateral bonds among Western nations is the fact that in peacetime they are in vigorous competition with each other (which in turn often translates into latent hostility among the populations), while in times of impending crisis or in time of war they are expected to be closely interdependent. The specific details of competition are important in the sense that their often burdensome impact is felt on a day-to-day basis by the populations and the governments, the very same which are being called upon to cooperate in support of common interests arguable only in the long-term perspective.   With very few exceptions, Western nations are importers of raw materials and exporters of consumer goods, capital equipment, and semi-finished products.   The U.S., Canada, Australia, and New Zealand are also agricultural exporters and in selected areas compete with each other and with members of the European Economic Community (EEC).   Canada, and perhaps eventually the United Kingdom, may remain or become self sufficient in energy supplies; all the other industrialized nations are (at least for the next few decades) vitally dependent on oil imported mostly from the Middle East.   All Western nations have aggressive domestic defense industries; they all wish to secure the economic fallout of defense business (including, whenever possible, the profits stemming from arms trade) and use the advanced technology derived from weapon development to enhance their competitiveness in the civilian marketplace.

The perception of future threats to societal institutions is widely divergent among the possible participants.   Europe has a long experience in dealing with diverse ideologies, including the presence of left-wing extremists.   Perhaps for this reason, anticommunism does not carry the same emotional overtones as those present in the U.S.   The attitude of many European nations toward socialist regimes is, to say the least, ambivalent.   Socialism and related regimes originated in France and England in the mid-19th century and spread from there toward Russia and the rest of the world.   A number of European regimes still retain a substantial minority (when not the majority) which is professedly socialist.   When added to the blandishments of the Soviets, expressed in the form of long-term trade agreements, including the supply of essential raw materials, such ambivalence may also offer tangible immediate benefits. [10] Let us not forget that Russia used to be historically the natural hinterland for Germany; for centuries the industrial surplus and know-how of Germany supplied the Russian economy.   In modern times Germany, as a result of World War II peace treaties, has the added incentive of the possible reunification with East Germany to suggest that its interests vis-a-vis Russia are not necessarily or permanently congruent with those of the United States.

[10]   (Note added in proof.) The current intense debate about the natural gas pipeline from Soviet Russia to West Germany is prototypical. A consortium of Western European banks is exploring the financial arrangements, with the active support of some U.S. banking institutions.

The traditional cleavage between the left and right in domestic politics is present at various depth within each potentially participant nation. The existence of Social Democrats vs. Christian Democrats in Germany, Communist vs. Gaullists in France, the equivalent in the British Labour and the Tory parties in England, with counterparts in Italy, Japan, and the Scandinavian nations, makes it clear that no democratic government can go all the way to one extreme or the other. The pendulum in matters of domestic politics oscillates back and forth. This, of course, renders implementation of any coherent long-term policy just as difficult as in the United States, which, ironically, for this reason is often subject to Allied criticism.

Western Europe and Japan certainly do not share the United States emphasis on the protection of Israel.[11] Our emotional commitment to the Israeli cause may have stronger roots in their strength within the American electorate than any abstract principle of rendering political entanglements permanent. We should not be surprised that our Allies do not share our conviction. Our recent policies in regard to Iran, Afghanistan, or Pakistan, our approach to reprisals against the Soviet Union, our attitudes toward restrictions on nuclear power[12] are not exactly conducive to the Allies being convinced about our competency, resolve, or the purity of our intentions. Many Western nations recognize the need for stringent oil and energy policies in the near term, but they hesitate to commit to (possibly nuclear) war over such an issue.[13] The European nations, and to some

[11]   Even the U.S. policies on the subject are somewhat ambivalent. If we say that the U.S. unconditionally stands behind its allies, then one should remember the relative ease whereby we cast off Taiwan, which was recognized for more than 20 years as the true government of China.

[12]   Interpreted by the British, French, German, and Japanese as a clumsy way of trying to retain hold on the nuclear power industry while using obsolete technology.

[13]   Let us not forget that Western Europe and Japan are emotionally scarred battlefields; they are economically most fragile, and they exhibit (if anything) even less resiliency in case of a major war than the United States because of the lack of relatively uninhabited hinterland. They have in the

extent the Japanese, fear the black-and-white approach of the U.S. to foreign policy problems which may embroil them in nuclear war with the Soviets without their having had full opportunity to choose their course in keeping with their national interests.

"Atlas and the Others"   *(Die Zeit)*

Many are the difficulties to progress in the direction of common political organization, but perhaps the biggest implementation problem is the absence of broadly accepted competent leadership. The era of the Churchills, Spaaks, and Monnets is long gone; very few statesmen speak for the

13 (Footnote continued)
    past benefited from the presence of the U.S. military shield and economic umbrella. They would go out of their way to retain such an umbrella but not at the risk of being dragged into war for unilaterally specified U.S. interests.

alliance as a whole and, frankly, the U.S. leadership has been found wanting.[14] The U.S. political system and quality of leadership has not been admired or fully accepted ever since the late Eisenhower years. The U.S. economic clout has weakened and the credibility of the nuclear umbrella is being increasingly questioned; they cannot therefore compensate for the loss of moral leadership. In point of fact, a number of U.S. political initiatives are discounted or are being seen as potentially damaging to Allied interests. The moral aspects of Watergate, the mistakes of Vietnam, and perhaps those in the Middle East, are often perceived as results of American lack of resolve and maturity.

In addition to all these elements of substance, there is also a matter of style. Each Western nation, owing to its own traditions or to the weaknesses already discussed in regard to the near-term/long-term dichotomy of democratic processes, has a tendency to muddle through problems just as does the U.S. Government. Long-range plans, long-term perceptions in the appropriate perspective are most unusual and decisions are invariably confused or postponed, unless stimulated by the immediate presence of an unmistakable threat.

*

A reasonable plan for implementation, evolved in cooperation with all potential participants, is an essential first step. It can be undertaken right now.

No matter how desirable the Commonwealth objectives may appear to the dispassionate analyst, the whole political concept would be ignored or rejected, unless a reasonable implementation plan is offered. The EEC, which is on the verge of reaching its 25th birthday, offers material for a valuable case study; most mistakes made during its inception phase might offer guidance for avoiding repetition.

---

[14] This is in sharp contrast with the process of agglomeration which took place back at the Albany Congress in 1756. At that time Benjamin Franklin, helped by a few friends of his junto, almost single-handedly put in writing what eventually became the principles of association (and eventually federation) among the then independent Colonies and "Plantations." Of course, community of interest and common language helped.

We just have observed that when approaching the implementation phase, even at the planning stage, it is quite likely that a number of "local" (e.g., national) objections will be raised.   After all, habits, attitudes, customs, vested interests, and political structures are often quite deeply ingrained and tend to be preserved for both emotional and economic reasons.   A new structure, such as the Commonwealth, should rest on the principle of what we might describe as the *democracy of national cultures.* Every individual group may and should, at its own option, retain its right to self determination in matters internal to that group.   The incremental implementation steps should ensure that no forced modifications of national ideals, ethnic or racial peculiarities, are being imposed as a consequence of their participation on those unwilling to accept them.

The supranational agglomeration should respect the principle of *stratification* of public powers. Stratification in this context means that each government level has the authority and the responsibility for solving only those problems which normally are within its purview.   Thus, the local, municipal, and county authorities in the United States handle most legal and regulatory matters which have to do with the day-to-day existence of the individual; the state law (or in other countries the provincial law) addresses such matters as trade, criminal justice, and, of course, taxation. At the national level, common defense, the authority to issue currency, and the postal service are typical.   The Commonwealth of Western nations at that "superfederal" level would deal exclusively with responsibilities *that have been expressly delegated to that level.*   The U.S. Constitution and history may offer examples and experiences of how these types of problems should be approached.

In general, especially at the initial stages, the Commonwealth should assume responsibility for "global" issues, such as population control, protection of the ecology, common financial and economic structure, strategic arms limitations, and more generally, negotiations with other major agglomerations such as the Soviet bloc and the Third World.   The Commonwealth authority should also sponsor joint policy formulation on matters of mutual interest such as international trade, labor laws, and migration policies. As experience develops, other relatively urgent and controversial problems may enter the sphere of Commonwealth responsibility.   Joint approaches to energy, raw materials, currency stabilization problems, and common policy toward nuclear weapon proliferation and arms trade outside of the Commonwealth are possible candidates.

The U.S. should not seek to impose its own leadership on the future Commonwealth. It has its own internal problems far from being fully resolved; it has lost its monetary or weapon clout, and its already questionable trade clout may be dwindling further.   The U.S. should behave as a responsible member of the Western community; it should show the example how to subordinate its narrow domestic interests to the broader goals of the community, and participate in the support of executive power delegated to the Commonwealth.   To the extent that U.S. leadership is deserved by future behavior, it should not imply dominance or dictation of long-term goals. It should support the step toward the establishment of the Commonwealth, and should as much as possible support measures of common defense in order to protect the process.

The U.S. has major ideological contributions to make to both the genesis and the actual operation of a Commonwealth type of supranational organization. We can offer considerable historical experience in operating a closely integrated pluralistic society, where authority is carefully designed and circumscribed by principle and usage to enforce explicit checks and balances. On the other hand, the U.S. Civil Service, as a whole, has considerable lack of background in foreign affairs. Traditionally such experience has resided in the private preserve of the State Department and cannot be easily transfused into other areas of the political process. Specifically the Department of Agriculture, Labor, and some of the major Congressional Committees have very little exposure to foreign policy issues, almost as a matter of principle.

The sequence of events which might lead to a gradual

establishment of such a Commonwealth is envisioned as follows: (1) Understanding of the arguments related to the establishment of a Commonwealth and general acceptance of the fact that, on balance, they are in its favor; (2) Plan and establish the principles of the initial steps; (3) Obtain legal status for the new authority and define the powers specifically delegated to the Commonwealth; (4) As successful operation is being experienced, open the door for new accretions.

Within the U.S. we should organize an official study group, including the present membership from the State Department, to expand and amplify the responsibilities of the Trilateral Commission under Government sponsorship. Related studies ought to be funded in the universities either by the Government or supported by foundations. A formal tie among the legislative bodies of potential participant countries ought to be established. Each country must share the costs and benefits with its Allies. In addition each major nation should retain military and economic reserves (including a fully competent industrial base) available to protect new accretions against potentially hostile initiatives.

A practical accretion procedure should be established which defines the standards and the methods of implementation as to how to have new nations be added to the Commonwealth. The pattern could be similar to the one used to establish statehood for new territories to become part of the United States. This might allow expansion of current alliance structure, such as NATO Europe, to eventually Japan, the Southeast Asian nations, the rest of the Western Hemisphere, and eventually to what is described as the ocean front, including most of the populous countries at the littoral of the great oceans of the world. Economic policy ought to be developed as a consequence in due course to support this alliance structure.

It should be abundantly clear that the process in establishing true and close cooperation leading to eventual political and military federation of the West as implied by the Commonwealth is a lengthy and tedious process. An interim common front for economic cooperation and a closely intertied common defense are necessary until the appropriate Commonwealth mechanisms are in place and a joint *modus vivendi* has developed toward the Second or Third Worlds.

*

Very few would, upon serious reflection, question the premise that the interests of the Western nations shared in common outweigh their differences or causes for competition. Once this premise has been accepted, even fewer would support the view that the current approach to pursue our common goals is purposeful and promising, let alone effective, in eliminating the risk of catastrophic failure. The area of disagreement is mostly focused on the desirability of supporting the common Western interests by an explicitly formulated supranational organization and, given this as a desirable objective, the pace and the nature of implementation steps leading to its achievement.

Even if the reader is not yet enthusiastically convinced at this point, the arguments in this chapter may have created at least a favorable prejudice in support of the Western Commonwealth idea. Assuming that eventually these arguments will be reexamined, expanded, and (I hope) much improved, the individual nations will enter a lengthy period during which their opinion-makers will slowly explore, and subsequently espouse, the concept; it may take a new generation before a significant majority among them will forcefully and consistently support a convergent approach.

In the meantime, those who are now convinced can substantially contribute to the rate of progress in the right direction. Steadfastness, patience, and dedication are, as usual in such matters, the essential ingredients; in this case I feel bound to add the desperate need for additional knowledge. Any critical examination of the line of reasoning given here reveals that it is essentially based on extrapolation of historical precedents and alleged analogies to a future period which, according to all premonitory signals, may be violently *nonstationary*.[15] The best brains of the political science community will have to be enlisted in order to define and promote the coalescence of incremental steps eventually leading to the Commonwealth.

If promising and offering reasonable evidence of growing success, the effort in this direction may well go down in history as the achievement which will have redeemed the sorry record of accidents and misjudgments of the twentieth century.

[15]   Once again, I borrow from my mathematician friends to describe a time sequence which cannot be predicted on basis of the statistical properties of its past history.

CHAPTER XII
Global Economic Structure

THIRD PREMISE: THE LONG-TERM STABILITY AND WELL-BEING OF WESTERN NATIONS, INDIVIDUALLY AND COLLECTIVELY, ARE CONTINGENT UPON THE DEVELOPMENT AND IMPLEMENTATION OF A CULTURAL AND ECONOMIC INTERACTION FRAMEWORK WITH THE LESS DEVELOPED NATIONS, ATTRACTIVE AND ACCEPTABLE TO ALL THE PARTICIPANTS.

While the Soviet Union is the major, immediate direct threat, the long-range problems faced by the West are largely, and perhaps decisively, determined by the fundamental competition with the less-developed countries. The trends now unleashed in the "Third World" are so irreversible in nature and of so momentous consequences that they would present a formidable challenge even in the absence of any further Soviet-initiated mischief.

This chapter deals with the problem of Western interaction with the less-developed countries (LDC's) in the next few decades. It attempts to make the points that such interaction is vital and unavoidable for the West; that it can be planned and structured to mutual advantage without significant disruptions; and that (provided the Western collective resolve and military strength remain adequate) the Soviet Union or other potential adversaries are not in position to interfere decisively with this process. By inference, the point is also made that the alternatives are far from being attractive, and that the West should rather urgently decide about its intended course of action.

*The development of stable, mutually beneficial interactions
with the less-developed nations is not a discretionary option
to the West; it is an imperative, urgent necessity.*

Interact we must—such is the proposition, but is it fundamentally true? [1] Is it possible to envision a world where the
West, having renounced the threat of military power as a
tool to enforce favorable terms of trade (and as a means for
justifying disregard of any aspirations evidenced by its
trading partners) would attempt to maintain as much as
possible of its previous "quality of life" under conditions of
more or less complete autarky? This is perhaps a feasible,
albeit not an attractive, option for the U.S.; it might be a
possible course for Canada, Australia, and New Zealand, but
certainly not for Japan or for Western Europe. All but the
latter two are net exporters of agricultural products as well
as of capital and consumer goods; while much of their trade
is with other industrialized countries, their economies would
suffer serious dislocations were they to be cut off from
their LDC markets and suppliers. In some (relatively few)
cases, essential "strategic" materials are mostly supplied by
LDC sources, but if given sufficient lead time, these needs
could be satisfied from mostly Western sources or by the use
of replacement materials.[2] For Japan and Western Europe,
autarky with respect to the LDC's is largely unthinkable;
their aggregate trade with the Third World is of the order of
23% of their total GNP (1978); this figure is projected to
grow to 27% by 1985.[4]1

[1]   Dependent on the degree of coercion assumed, one can imagine a wide
range of interaction models. With the harsh exercise of world-wide military
power (not exactly an inexpensive undertaking), it may be possible to
maintain the per capita wealth ratio of the industrialized nations to the
LCD's between 12 and 15 (bracketing the range of estimates in regard to
current values). It is also possible to engage in the purely theoretical speculation that the Western governments would be successful in persuading
their respective electorates that their fair share of the planet's wealth is not
more than (say) 10 to 15% above the world-wide average.

[2]   For almost all Western nations, oil at affordable prices is considered vital
to the point where it is seriously suggested that nuclear war with the
Soviet Union is an acceptable risk in order to protect its continuing availability. It is not necessary to endorse this view in order to gather an indica-
tion of the importance of oil supplies in the judgment of many. In this

The integrity of the West as a group of nations is jeopard-ized by the combination of military threats and economic perturbations inspired and abetted by the Soviet Union, as well as by the erosion (equally abetted by the Soviets) of their cohesion and will to fight for survival. If the West elects to isolate itself from the rest of the less-developed world, it will, as a group, considerably decrease its own growth rate, which in turn will continue to erode its standard of living and to cause further increases in both the incidence and the intensity of social stresses. Having deprived itself from its traditional source of growth and of disposable surplus, the West will no longer be in position, or in the mood, to support the ideological and military commit-ments required to face the vigorous direct competition by the Soviets. In the more distant perspective, the same holds true against other, perhaps just as dangerous, challenges by newly emerging competitors.

There is another facet of the case for continuing, stable, interdependence between the West and the LDC's. Owing to their own internal social and economic divergencies, the developing world faces an era of instability and growing stresses [3] which, in the absence of mutually beneficial Western involvement, may lead to serious, not to say catastrophic, dislocations. This would still further delay the LDC's emancipation as self-sufficient, stable nations and at the same time offer attractive opportunities for the Soviets

2 (Footnote continued)

connection, I would like to advance, as a purely personal conviction, that the risks of serious military confrontations (including nuclear war) with the Soviet Union are *entirely determined by our overall strategic posture* in relation to the potential adversaries. If we are unable to convince the Soviets that they have nothing to gain, and much to lose, from provoking war, then, at that juncture, the exact cause of, or pretense for, the con-frontation is almost immaterial. On the other hand, even a relatively mild nuclear war (if one may be permitted to even think of such a word in such a context) would cause economic disruptions far in excess of any possible repercussions resulting from a complete oil import cutoff. This is emphat-ically true for our Allies; the thought that the U.S. and the Soviets are engaged in war over the Middle East oil supplies while the Allies are enjoy-ing business as usual in the midst of plentiful oil supplies is, to put it mildly, unrealistic.

3 Appendix—Note S.

to exploit the recurrent crisis situations to their own cumulative advantage. Should the West elect to isolate itself culturally, economically, and militarily from the LDC's, they will not remain free to choose a path of peaceful, long-term evolution, but will be in turn enticed, exploited, subverted, and eventually subjugated by our less scrupulous, but more aggressive, competitors. The Soviet-advocated theory of "socialist correlation of forces" would receive thereby further powerful corroboration and reinforcement through the Western inability of perceiving, and of deliberately furthering, its own interests, even when these are patently immediate and potentially vital.

An ethical component is obviously present in the increased concern among the Western nations for the future of the less-favored segment of the developing nations. While resentment is strong against the OPEC cartel and similar manifestations (for some strange reason interpreted as con-spiracies against, and ingratitude toward, the Western world rather than the hard-nosed economic "realpolitik" which would be so highly regarded if pursued by one of our Western colleagues, or even better, by ourselves), the development of social consciousness and the racial affinities among our own minorities prompt us to apply a measure of human understanding in our relationships with the "have-not" nations.[4]

Failure to establish a long-term, valid framework, within which the resources and the wealth of the planet can be shared on the basis of some peaceful understanding of the concept of "equity" (more about this later) would have some further ominous future consequences, quite independently of any Soviet interference:

- In some cases, the very rate of population growth, coupled with the highly visible differences in standards of living may create sources of tension and eventually conflict. Such "discontinuities" (as distinguished from

[4] This is one of the reasons why food supplies are, in general, not denied on political grounds, even to countries manifestly violating all rules of civilized diplomatic intercourse.

gradients) across national boundaries are now present across the U.S./Mexican border and the U.S./Caribbean littoral; the lines between India, Southeast Asia, Indonesia, Indochina on one hand, Japan and Australia on the other, may soon be exhibiting the same phenomenon.

- In the much more distant future, some of the major population foci (China, India, Brazil, Mexico) may, through internal development or through coalition with others, acquire the leadership and the cohesion necessary to pose a real threat to world stability. A nuclear-armed China or India is not so far fetched; if spurred on by power on one hand and by the specter of deprivation on the other, wars may be the traditional remedy; a prospect by no means attractive to the West.

*

Instances of success in Western interactions with the less developed countries are at best sporadic and the results ephemeral.

Surveying the history of Western dealings with the LDC's since the mid-1950's this assertion may well appear quite an understatement. Granted that the secular changes in the political environment [5] and the bitter residuals of past conflicts render some of the associated problems almost intractable, it is fair to observe that the Western nations collectively have failed to bring to bear the requisite insight, wisdom, and sense of purpose on the efforts aimed at their resolution.

The legacies of the past do not affect all Western nations in the same manner or to the same extent. Both the United Kingdom and France have managed to retain significant cultural and economic ties with their former colonies and dominions;[42,43] The Netherlands has also preserved its preferred position in regard to Indonesia. The roles of Belgium in regard to the Congo, and that of Portugal to its former African possessions, are more ambiguous. Germany

---

[5] Chapter VII.

has not been a colonial power since 1919, and as to Japan, its current regime appears not to suffer overtly from the stigma of colonialism in what used to be called the "co-prosperity sphere."[6]

Paradoxically, it is the U.S., which for practical purposes never possessed major colonies[7] and voluntarily relinquished those it still held at the close of World War II which is made to suffer the most the burden of past exploitations. Following the Second World War, the U.S. was seen as the richest and the most powerful nation in the world. Its power was visible everywhere and its economic domination was almost complete. Following the open split with Russia (c.1947), the U.S. also chose to underwrite the safety of the Free World from the threat of communism. Following our mercantile interests and under the guise of multinationals, the U.S. of necessity took an ambivalent position: the Government officially was liberal and against colonial exploitation, but in everyday transactions the U.S. often sided with the forces of repression. Soon, with the help of Soviet propaganda, the U.S. was held up as the villain of practically all the struggles in the wars of anticolonialism and of national liberation. The unfortunate sequels of the Vietnam involvement, compounded by our fiascos or abstensions in Angola, Somaliland, Ethiopia, Cuba, and most recently in Iran, further contribute to this impression.

The record of the West, and more particularly that of the U.S. in the past 15 years or so, is little short of dismal. Outcomes in specific instances of crisis seldom favor the West; in the rare cases when success can be plausibly argued, the results are often short-lived or purchased by expensive concessions in other areas of concern:

[6]   Taiwan (formerly Formosa), Malaysia, Singapore, French Indochina, parts of Indonesia. Manchuria is now fully part of the Peoples Republic of China.

[7]   The Philippines gained independence in 1974; Okinawa was not a U.S. colony per se, but was nonetheless restituted to Japan in 1971. Hawaii acceded to statehood in 1960, and Puerto-Rico gained Commonwealth status in 1964.

- The Western governments have failed to perceive the rise and the growing cohesion of the OPEC. The impact of the 1973 oil embargo forced the U.S. into an "even-handed" Middle East policy at the expense of Israel. While this has secured the newly evolved rapport with Egypt, the Camp David accord failed to address (let alone resolve) the crucial Palestinian problem. It has equally failed to gain endorsement by the Arabs who really count: the conservative petropowers (Saudi Arabia and Kuwait) and the radical confrontation states (Iraq, Syria, Lybia, and Algeria).

- The recent history of U.S. relations with Iran is still too fresh in our memory to require much elaboration.   In gross terms, though, after publicly supporting the late Shah's regime for more than a quarter of a century, our ostensible friendship and much publicized military power were unable to protect the regime, or even our own obvious interests, against a clearly predictable internal upheaval.   Our own hesitancies, between favoring military power and insisting on increased human rights according to U.S. standards, strongly contributed to the swiftness of the Shah's demise and to the strong anti-American strain detectable at the core of the new Islamic Republic.

- To the extent that any coherent U.S. policy had been defined and pursued in regard to Black Africa, it has been subject to reversal several times in the past decade.   In the last Nixon years, it was "business as usual" with South Africa; the clamors of the Black African nations and the strident denunciations of the U.S. black minorities were quietly ignored. So were, in all fairness, the disturbing rumors about the excesses of Bokassa in the Congo and of Idi Amin in Uganda. Rhodesia and Southwest Africa (a.k.a. Zimbabwe and Namibia, respectively and recently) were problems left to be handled by those directly involved.   In the very last   few months of the Ford Administration, Dr. Kissinger announced a sweeping change in U.S. policy: thenceforth, the U.S. will recognize the legitimacy of black aspirations over the African continent and will dissociate itself from the racist policies of the remaining white-dominated nations. [8] This trend was reinforced during the Carter Administration, especially with Andrew Young, U.S. ambassador to the U.N., carrying the same message forcefully, if not always diplomatically, to all those who wanted to listen within the U.S. or abroad.   The facts that South Africa is a respectable trading partner; that it is the principal source of gold, diamonds, and of some minerals of considerable strategic   importance; that it is located athwart of essential maritime communication lines; have   been   conveniently   ignored   or   at   least deemphasized.   There are reasons to suspect that with the new Republican Administration under President Reagan, this attitude will be at least reassessed, and possibly reversed.   In view of these frequent and significant shifts in policy, it is not surprising that the U.S. is not exactly perceived as a reliable, steadfast ally by anyone in these parts; nor is it clear that a more coherent   or   more   effective   regional   policy   is   being evolved.

• The long debate on the Panama Canal treaty has finally resulted in the virtual cessation of our legal claims to the Canal ownership (by the end of this century, with appropriate safeguards for security and passage rights); it was presented to the American public as the major step to win respect and friendship in Latin America, and more particularly in the Caribbean.   It is regrettable to observe that even the most superficial perusal of the daily news, reporting the dismal concatenation of events   of   Nicaragua,   Guatemala,   and   now   of   El Salvador, suggests that our recent efforts of good will turned out in some mysterious way to be misguided or at   least   insufficient   in   the   face   of   internal   social stresses, probably abetted by the Castro regime's efforts to export its own revolution.

• In other parts of Latin America, President Carter's human   rights   policy   may   have   impressed   our   own

[8] Portugal relinquished its African possessions in 1974.

domestic liberals, but it certainly has not gained favor with the "conservative' military dominated regimes of Argentina, Brazil, and Chile. It has obviously not impressed at all the leftist opppositions and the large masses of "underprivileged" in these countries, to the extent that these have any voice in policy formulation.

This enumeration could easily be continued to cover the problems which loom on the horizon in our relations with Mexico, India, and Indonesia—all nations with large or rapidly growing populations which are expected to play important roles in 15 or 20 years; all are essentially ignored in our long-term policy formulation until some immediate crisis draws attention. Perhaps a fair summary would be that our unqualified successes are difficult to identify, while uneasy compromises or continuing delays and vacillations are by far more characteristic of our relationships to the less-developed nations.

Based on the preceding chapters, there should be no particular difficulty in identifying the causes of the collective inability of the West to come to grips with the problems facing them. (I) At the present time, in many international endeavors the Western nations still act at cross-purposes. Each pursues its own relatively short-term interests as derived from tradition or as imposed by its own domestic perceptions; cooperation, especially when it calls for sacrifices, is the exception and by no means the norm. The Western nations compete with each other for raw materials, markets, and investment opportunities; their LDC counterparts exploit this competition to drive their shrewd bargains to the point where the trading advantages become marginal for the West. [9] Not so incidentally, the industrialized Western nations are no longer the sole, or even the principal, source of investment capital; the voices of the capital-rich petropowers are increasingly audible in the

---

[9] Trading advantages are questionable when the West must expend its scarce capital to assuage its insatiable thirst for oil, or where it attempts to gain some influence (short lived, in general) by introducing new types of weapons into a given region.

hallowed precincts of international finance. (2) In most instances, the West still regards relationships with the less developed countries as primarily an economic interaction, rather than a broadly based geopolitical problem with long-term implications. Policy formulation in this area is still largely influenced, not to say determined, by the large trading companies or by their modern descendants. It is somewhat naive to suppose that the majority of these mercantile interests will act in an enlightened manner to promote the long-term interests of the host nations, although some gratifying counterexamples can be found.[10] Until very recently the public at large in the West and their elected representatives or governments have tended to perceive the LDC's as interesting areas for tourism, or deserving subjects for charitable support when stricken by drought, floods, earthquakes or epidemics. That the less developed countries have national aspirations, strongly held traditions, and serious clout in international negotiations are all concepts which enter the Western world view only very slowly, and even then under the pressure of some immediately felt unpleasantness. (3) The U.S. behavior, as contrasted to its Western partners, is particularly subjected to confusion or even ambivalences bordering on the edge of schizophrenia. Many instances can be cited where we are torn between anticolonialism and economic interests, anticommunism and human rights,[11] foreign policy constraints and domestic minority or industrial pressures. (4) The rationality in the behavior of our potential LDC partners can not be judged by traditional Western standards. Their viewpoints are influenced (distorted, in our eyes) by the legacies of their own past, so different from ours; by their own internal divergencies due to tribal or racial

---

[10]   The IBM Company promotes genuine in-country research and personnel development in places as diverse as Western Europe, South America, and India.

[11]   The two "showcase" countries, the Republic of Korea and the Republic of China (a.k.a. Taiwan) show the remarkable effects of prosperity, Western style, but they cannot be considered as paragons of behavior in regard to democratic freedoms. Our unilateral withdrawal of diplomatic recognition from Taiwan, our erstwhile ally, could not remain unnoticed by other nations in the Far East and elsewhere.

causes; by their relatively recent interpenetration with the Western societies and by the cautious, when not hostile, attitudes toward anything reminding them of their colonial periods.   As the LDC societies evolve, and often as a consequence of rapidly   changing economic conditions or cultural differences, internal stratifications and divergencies appear and are reflected in their national attitudes and behavior. (5) Lest we forget, as pointed out earlier, for the near-term future the Soviet Union and its proxies are present and ready at just about any opportunity to exploit misunderstandings or crisis situations to their putative advantage.

It should be obvious, with all these impediments, that satisfactory approaches to evolving a stable cultural and economic framework between the West and the Third World can not be conceived without identifying the long-term bases for such interactions, which should (hopefully) remain unchanged and predominant with respect to all the foregoing.

* * *

The long-term interactions of the West with the (now) less-developed nations are dominated by their inherent characteristics, not by the designs or actions of the Soviet Union.

It is instructive to speculate for a brief moment on how the future course of history would be modified if all of a sudden the Soviet Union would drop out from the international competition.   Imagine that every member of the Soviet Communist Party, every government official (say, above the rank of GS-14 equivalent), every commissioned military officer above the rank of captain would, each and all, resign their positions, renounce power, wealth, privilege, and decide to follow Solzhenitsyn toward the building of a virtuous, nonaggressive, and spiritually motivated Christian Russia.   The probability of such a sequence of events is admittedly low; it is nonetheless a fascinating subject for reflection.

The U.S. would once again enjoy the undisputed supremacy in nuclear weapons. The British would have scrapped theirs for reasons of economy; the French would have mothballed most of theirs and sold the rest (in questionable state of readiness) to India. The PRC would continue its own nuclear weapons program, but at a relatively low level.

The U.S. Congress would feel justified in slashing the defense budget by 60%; most of the large hardware procurements are stretched out to the point where they become for all practical purposes WPA programs for highly skilled, but virtually unemployable, technical teams. In a fit of virtuous indignation, Congress even proceeds to eliminate large chunks of veteran benefits and moves to make military pensions retroactively unattractive. Many large U.S. defense contractors "diversify," i.e., look for greener pastures, while waiting for an upturn in the traditional aerospace business. Another series of long, hot summers, interspersed with disastrous, but nowhere unusual, sequence of frigid winters, devastating floods, hurricanes, and transportation tie-ups bring home the frightening truth that the procurement portion of the U.S. military expenditures is really a miniscule fraction of the U.S. Government budget; that whatever revenue can be possibly raised will be invariably committed to purposes perhaps worthwhile but perceived as being absolutely vital. Inflation, unemployment, and unrest would continue within the U.S. and within most of the West.

It is certainly possible, but not very likely, that the West under such circumstances would retain its cohesive military power.[12] It is also possible, but not very likely, that a large proportion of our younger generation will once again acquire the adventuresome pioneering spirit and volunteer enthusiastically for overseas duty involving hardship and danger in exchange for glory and (the modern equivalent of) plunder. It is far more likely that we will manage to retain only a relatively small core of military power effectively

---

[12]   In the next chapter, we will argue that even with the massive Soviet threat clearly present, it is not easy to motivate the post-industrial West to maintain powerful military forces in a permanent state of readiness.

usable overseas. It is also quite probable that the competition among Western nations will in fact neutralize attempts to dominate the less-developed portions of the globe in some newly defined relationship based on power. As for the U.S. in particular, once the Soviet threat is no longer perceived, our resolve in favor of military domination would be considerably weakened by the moral and ideological ambivalences mentioned a short while ago.

Irrespective of our ambitions and intentions, the nature of the potential target LDC's will have changed to a surprising extent; it is hardly possible that this change would be reversed. None of the current LDC's will have within the next 25 years the power to effectively challenge the coalition of Western powers (if such coalition exists), but they will possess certainly the forces and the motivation to resist vigorously any encroachment upon their sovereignty. Overseas expeditions of conquest would no longer be glorious and inexpensive adventures for the West, in fact they have never been since the late nineteenth century; they are far more likely to turn into unrewarding sacrifices of life and treasure. This likelihood would be even stronger if the intra-Western competition leads to a clash of spheres of influence. With the Soviet Union out of the picture, it is perfectly possible to imagine Japan in a more aggressive role trying to carve out for herself a new co-prosperity sphere, perferably including some of the choice sources for oil, minerals, and agricultural products.

The best we can really hope is that we may acquire, through negotiations subliminally backed by power, a few strategic bases around the world (e.g., Persian Gulf, Malaysia, Caribbean) for the purpose of effective presence that may deter irresponsible support of international terrorism or local acts of aggression. Collective help against "subversion" is possible but not likely to be very effective; as we pointed out earlier, perceptions about subversion are largely a matter of local perspective. One possibility is collective military action against the proliferation of nuclear weapons, but even this is not highly promising in view of the many surreptitous moves which are likely to surround nuclear developments in the future.

The central point of all this is that the basic nature of the long-term relationships with the Third World will not have changed for the West just because of the hypothetical fact that the Soviet Union has ceased to be an effective military threat. The current role of the Soviets should be seen as accelerating the LDC evolution toward independence by neutralizing some of the potential Western military initiatives and by attempting to counter the attractiveness of Western ideologies. The Soviets are not the *cause* of the evolution since 1945; they simply exploit and exacerbate the causes which are intrinsic to the nature of the LDC's. We in fact encouraged or condoned this evolution in the past so that its effects by now have become essentially irreversible.

The causes of stresses in the Western societies have earlier been identified as resulting from considerations mostly internal to each nation (information flow and perception of social benefits); and from the combined effects of ecological and resource constraints compounded by the economic pressures resulting from the increasingly assertive positions assumed by the Third World suppliers. There are many details and complicating factors; no doubt the apprehension of growing Soviet power contributes to the general feeling of pressure. It should be evident, however, that (assuming the relatively near-term Soviet threat to be effectively neutralized) the fundamental problem which faces the West in the long term is the competition with the less-developed countries of the world for power and influence. In more immediate terms, this means access to the sources of wealth available on the planet: natural resources, agricultural products, human resources (including labor, skills, and culture), technology, capital, and management knowhow.

This long-term competition at the global scale is without precedent in world history. The past political and economic structure, backed by Western military power and justified by technological aspects of cultural differences, has been mostly destroyed *de jure* and essentially weakened *de facto*.[13] The presence and rapid growth of the LDC's are a

13 Chapter VII.

fact of life in the late 20th century which we must accept and learn to accommodate. There is no returning to the "good" old times when famine, pestilence, floods, and wars were considered routine occurrences, fit subjects for polite conversation, but not otherwise objectionable. Today, with the progress of medical science and the spread of even rudimentary health care, the population growth of the LDC's will not slow down significantly in the next 20 to 25 years. The political, economic, and social divergencies within and among the LDC's are rapidly growing and offer ample opportunities for disruptions and upheavals with the ominous long-term component, the relentless growth in the biologically aggressive population segments, everpresent as a frightening backdrop. Soviet efforts at securing advantage from deliberately provoked crises contribute to the volatility of the situation, but they are hardly essential.[14]

For better or worse, the West must, in the long run, find some way to share the resources available [15] with its communist competitors and with its Third World shipmates on "Spaceship Earth." The results of analyses, purporting to show that, according to some "lifeboat theory"[44], the world would be better off by allowing the have-not nations to be decimated by famine before they waste (sic) valuable food and fuel resources, are (understandably) not accepted by the potential candidates for starvation. Such is our perverse morality that they are not even seriously considered by the well-appointed seekers of truth in the groves of academe and of the not-for-profit think-tanks of the West. Whatever sources of food and of other means of sustenance or enjoyment can be exploited will have to be in some way shared by all the inhabitants of this planet; only the proportions and the mechanisms involved in this sharing are possible subjects of argument. But none of these will remain permanent, none of these will be satisfactory to all to the point where strife and contention might be eliminated.

---

[14] Appendix—Note S.

[15] More precisely, exploitable at economically rewarding cost.

Let us examine now whether or not it is possible to conceive and to implement any specific course of action, compatible with the currently extant facts of international life, which would offer a reasonable promise of relatively stress-free evolution in the LDC's relationships to the industrialized West.

*

The deliberate creation and fostering of multilateral, dynamic gradients, perceived by all as culturally and economically attractive, are the key elements of the stable future interaction framework.

The central thesis of this chapter is that a stable interaction framework between the West and the less developed countries (LDC's) can be evolved and implemented in spite of all the currently perceived impediments; and that such a framework is possible only if the *gradients* which undergird and drive the future interactions remain valid.

In an earlier chapter, the concept of gradients was introduced. They are cultural and economic differences, in general covariant with separation and distance, which result in assignment of different local values to products and commodities forming the potential basis of trade between two or more societal groups. The intrinsic reasons for the existence of gradients may be real (e.g., land, minerals, industrial processes) or purely perceptual (decorations, drugs, tourism), but certain characteristics must be present. The substance traded must be abundant and cheap in the place of export; it must be scarce and highly desirable (and therefore expensive) in the place of import. Unilateral gradients are useless (they do not lead to stable trade relationships) or even worse, they usually lead to violence in the absence of countervailing trade. Bilateral gradients, (i.e., when two potential trading partners enjoy opposite gradients with respect to different but roughly equivalent commodities) are conducive to stable trade relationships, provided that they remain truly bilateral and equivalent. Multilateral gradients are possible and frequent; they may well be the wave of the future. Whether or not trade driven by a given gradient pair is in the long-term interest of those who enjoy its near-term benefits is an open question; the answer depends on the correctness of the participants' value judgments. In the long distant past, communications were so infrequent and contacts so precarious that only the most obvious gradients were exploited. On the other hand, the rate of social evolution on all sides was relatively slow, so that trading patterns could be established to last several decades and sometimes centuries.

The basic reason for the recent destabilization of world economy is the relative weakening of the once powerful bilateral trade gradients and simultaneous disappearance of the military power which would have maintained the economic linkages even in the absence of true gradients. These recent trends have been discussed at length in Chapter VII; all that needs to be added here is that unless they are drastically reversed, the chances of peaceful stable relationships between the industralized West (sometimes referred to as the "North") and the less developed countries (sometimes, even less accurately, referred to as the "South") are very slender indeed. Let us now examine in more detail the proposition that the creation, the strengthening, and the careful fostering of gradients are highly desirable and in fact indispensable if such stable and peaceful relationships are to be achieved.

First and foremost, gradients are beneficial because they create wealth. The potential benefits (not necessarily or always measured in profits) stimulate ingenuity, prompt the exploration and the exploitation of natural resources, enlist and energize skills and labor, attract investment capital; the benefits, once realized, more or less efficiently perfuse all segments of the participating societies. In the context discussed here, gradients require and promote communications between societal groups which otherwise would remain culturally isolated, even though competing in the ecological sense for the same resources. Rather sooner than later, their interests are likely to collide[16] and unless by then

---

[16]   E.g., grazing lands, hunting grounds, fisheries; but many other contemporary examples may be cited, such as environmental pollution across national boundaries, exploitation of deep-sea bed minerals, etc.

common value judgments and joint vested interests in the *status quo* are present, such conflicts are likely to result in violent confrontations.   By the same token, increased communications and shared value judgments contribute to the ability to withstand internal and accidental perturbations or deliberate provocations by parties outside of the trading group.

Participating in international trade and in the concomitant internationally competitive development process is the only hope for the LDC's if they are to realize their full economic and social potential in the foreseeable future.   Perhaps it is emotionally attractive to mouth such slogans as "global redistribution of wealth"; it is even possible that some symbolic gestures by the group of industrially advanced nations will be made in that direction.   But among the more knowledgable rulers of the LDC's, the apperception is very clear that their only real disposable wealth will not be received as a gift from others but created by their own efforts.   The examples of Japan and Switzerland show what can be accomplished by nations with practically no other natural resource than their industrious, disciplined and highly educated population.

Potentially, the gradients are present; the most obviously attractive ones are exploited to some extent.   They can be augmented and intensified by careful exploitation of other opportunities.   They will not remain valid for very long, however, unless carefully nurtured and continuously developed by deliberate and well-planned investment.   In other words, gradients do exist now, due to the large disparities in the resources and the characteristics of the different societies, but as these societies develop at different rates and in possibly convergent directions, the gradients tend to weaken and even to disappear.[17]   The

---

17   Thus, the traditional picture is for a less-developed country to export its (say) mineral resources to its industrial trading partner in exchange for industrial and consumer products, such as trucks and automobiles. As the LDC society matures, it would tend to request and to install truck factories of its own, even if in the initial stages, it cannot compete on the world market. Very soon it will compete with the original source of management and manufacturing knowhow, first in its own domestic market, while still

deliberate and continuous efforts to maintain and to expand trade gradients are an essential part of the global economic framework. It is a cooperative multilateral undertaking to improve and shift the export and import capabilities of all participants in order to perpetuate a stable, balanced gradient structure. It results in a steadily growing trade volume, flexible in its specific details but firmly oriented toward long-term mutual benefits. This is the condition described as *dynamic* equilibrium of the gradient structure.

Perhaps the most successful example of the deliberate creation of dynamic gradients is that of post-war Japan. Originally known as exporters of textiles and relatively low-quality industrial products, Japan has successively tackled photographic equipment, shipbuilding, electronics (and lately, microelectronics), and, of course, automobiles. In the last two areas, the gradients in fact have been reversed in regard to the U.S. and are quite robust in regard to all other nations, including the LDC's. In regard to ship-building, after having capitalized on the technical revolution of container ships and VLCC's, Japan has subsequently exported that technology, by means of investments and joint ventures, to Korea and Taiwan, freeing thereby its own skilled labor for other, more profitable purposes. At the same time, Japan has been aggressively pursuing a program of capital investment and technology transfer toward South America, Indonesia, and (surprisingly) Australia. In these areas, the primary objective is to develop and to diversify sources of supply of fuel, mineral resources, and other raw materials. As a valuable by-product, Japan also captures by this approach a substantial export market for its heavy industry aimed at transportation, utilities, and manufacturing processes. These latter used to be the preserve of the U.S. and of Western Europe for most of the first half of the present century.

---

17 (Footnote continued)
    importing for a while machine tools, spare parts, and designs. Ultimately, the newly developed manufacturing capability will compete with the original source in the world markets. At that point, the particular gradient will have vanished.

In spite of the well-documented success of some major trade relationships in the recent past, the Western powers, including the U.S., can not rest on the assumption that their societies will automatically generate and perpetuate the wherewithal for continuing profitable gradients. As the industrial nations mature, their production plants require continual capitalization just to remain competitive. Their attempts to protect their environment and the rising social expenditures exhaust (when they do not exceed) their disposable surplus of capital and technology resources. At the other end, the LDC's have learned the advantages of husbanding their natural resources; they face increasing burdens of capitalization and they manifest increasing ambitions toward developing more sophisticated industrial capabilities. All potential participants must seek out those gradients which are now profitable and attractive; they must continually invest in maintaining their strength and viability. In addition, they must seek in a cooperative effort to identify potentially *new gradients* to take the place when the traditional endeavors fall by the wayside. The maintenance of gradients in general depends on (1) The clear perception of evolutionary trends on all sides; (2) The understanding of the mechanisms which may perpetuate or supplant the gradients which may exhibit signs of weakening; and (3) The deliberate and purposeful investment in generating, accelerating, or reinforcing the new, long-term oriented elements. Even in the absence of explicit military coercion, the volume of exchanges between the industrialized and less-developed nations has increased manifold in the post-World War II decades, with the commendable initiatives of private multinational companies and with the benevolent (if sometimes counterproductive) attention of the governments. It is reasonable to assume that a concerted long-term oriented effort, basically involving the private sectors but stimulated, guided and supported by the governments, would result in a most rewarding and robust gradient structure which may form the basis of the global economic framework advocated here.

In further support of the thought that the deliberate fostering of a stronger, long-term oriented gradient struc-ture may be rewarding, let us observe that the nature of the trade substance which leads to the establishment of gradients appears to change. Instead of single products or commodities, we have come to consider broad multi-product trade agreements encompassing a significant fraction of the potential exchanges between partners. As our perceptions of the mutual needs evolve, it is perhaps not too farfetched to suggest that in the future trading partners may consider their respective long-term societal needs and then mobilize and coordinate their actions to cater to them. Instead of using the profitability (in the strict accounting sense) as the sole yardstick to determine whether or not a specific exchange ought to take place, the full range of interacting transactions, some of them showing benefits in the much longer term perspective only, could (and indeed should) be considered as the "substance" of the trade.[18]

The range of possible interactions is seen as increasing at a very rapid rate. Long past are the times when only physical goods and commodities are being traded. A very superficially compiled list would read as follows:

- Invisible exports (tourism, transportation, insurance)
- Economic services (investments, commodity options)
- Marketing and distribution support
- Education, training, and communications
- Administration, organization, and management training
- Medical and health care delivery; specifically including family planning
- City planning, housing, infrastructure development
- Agricultural development, including botanical and genetic research
- Transportation
- Exploration for natural resources

---

[18] This is happening right now, in more instances than commonly believed. It is instructive to note that some of the petropowers and LDC's are in the forefront of the trend. Saudi Arabia, for instance, has devoted considerable attention to identify policies and actions which ensure the stability of the Western economic structure. Granted that by doing so they pursue their own interest as well, but this, of course, is the basis of "bilateral" (mutually beneficial) gradients.

Many of these come under the rubric of "services" and have only recently entered the domain of deliberately "traded" components of international exchanges.   Let us keep in mind, however, that the "services" category constitutes a rapidly increasing portion of the industrialized nations' GNP; in the U.S. it already exceeds 50% and as the LDC's advance toward industrialization, the proportion of services required by their increasingly sophisticated societies will grow as well.   Extending the trade gradients in the direction of services (or to be more precise, a mix of services and products or commodities) increases the mutual opportunity to gain knowledge about each other's language, civilization, and aspirations.   As noted before, such knowledge, if properly applied, may go a long way in the direction of attenuating social stresses and of eliminating the potential causes of conflict.

For the foreseeable future, the differences between LDC's and the industrialized or postindustrial societies are so broad and so fundamental that they may be safely relied upon to provide ample gradient development opportunities. What, furthermore, is essential to observe is that these potential gradients almost invariably favor the West. Given that development (at the pace determined by the nature of their societies) is the true basis for stable growth and well-being of the LDC's, the industrial nations, together with the petropowers, are the only source of exportable technology, management know-how, investment capital, and food surplus at the scale required by the Third and Fourth World nations. At the same time, the industrialized nations, as a group, constitute for all practical purposes the only real export market for commodities, raw materials, and low-skill, labor-intensive products.   Together, these factors define the large-scale, intrinsic, bilateral gradients which should justify the confidence expressed here that a stable inter-national framework is not only desirable but indeed possible.

If the foregoing is true, then the "correlation of forces" so highly advertised by the Soviets, should really be in the favor of the West rather than in favor of the Soviet-inspired world view.   If our cultural and economic strength is properly marshalled to support the long-term interests of the less-developed world, then conceivably the LDC's will find little attraction in importing socio-economic systems offered or imposed by ideological considerations only.[19] Following the turbulence associated with the decolonization process of the 1960's, the revolutionary rhetoric, attempting to cling to the classical (and mostly irrelevant) slogans of marxism, was soon replaced by a more pragmatic approach of the newly emerging nations to the pursuit of their political and economic endeavors.[45,46] It is incumbent upon the Western policy makers to capitalize on this trend.

*

Many and serious impediments are to be encountered in the process leading to the stable global economic structure, but none of them are insuperable.

Even if one is fully convinced that the aggressive effort toward maintaining the multilateral gradient structure is justified, it remains to be shown that such an effort can be pursued at the implementation level with reasonable hope of success.   Assuming that the goals are accepted, in other words, what specifically should we do differently from what we are doing now?   As we have found several times in the course of this study, the general answer seems to be that we do not really have to modify radically our views or our behavior; the change is more a matter of apperception and of long range conscious commitment to reasonably well-defined objectives. More than that, it is a matter of judging systematically the appropriateness of our actions in refer-ence to these objectives.   Can we indeed formulate and often constrain our near-term actions by keeping in mind the more diffuse long-term goals?   We have argued earlier that this is not always (some would say hardly ever) possible in regard to domestic situations.

---

19   The Soviet Union is dominated by the *White* Russians; let us not forget the fact that they are seen by African and Asian populations as industrialized and relatively wealthy potential oppressors. The stern societal discipline advocated by the communist doctrine is not viewed with particular favor by populations not enured to the rigors of the Russian past.

It is reasonable to start with the need to eliminate or to attenuate the more blatant sources of immediate difficulties. (1)   The Western nations should define their common objectives and develop the mechanisms which allow them to act in concert on matters related to political and economic relations to the Third World.   Such cooperation is one of the basic purposes of the commonwealth-type supranational entity described and advocated in the preceding chapter.   (2) The Western public and its governments should cease to regard the LDC's as an irrational bunch of hotheads attempting to exploit or to punish the West, or as naive potential victims ready to fall into the deadly embrace of communism, or again as somewhat obstreperous suppliers of raw materials who can be cowed or enticed into submission by a show of force, edulcorated perhaps with a dash of bribery. (3) The U.S. should publicize and advocate its own political and social principles, but should refrain from forcing them on the potential trading partners.   We should not attempt to use our own standards in regard to human rights, labor legislation, nondiscrimination and many other areas of   social endeavors as a yardstick to determine whether or not a given nation "deserves" the blessings of U.S. trade.   This is not to say that we should condone or encourage clearly repressive regimes, it simply   suggests that our partners should be allowed and encouraged to progress toward their own goals at their own pace, rather than being forced to espouse our goals and accept our pace. Doing otherwise has proved foolish and unrewarding in the past; there is no reason to assume that it would be different in the future, especially not toward nations which by then will have developed potent aspirations of their own. (4) We should not delude ourselves into thinking that supplying weapons in exchange for trade would form a dependable basis for beneficial interactions in the longer term perspective.   The reasons were given earlier; suffice to repeat here that commitment to mutual security, including weapons and possible military presence, is a valuable export commodity indeed, but only to nations and regimes which have over a period of several years demonstrated their steadfast adherence to Western goals and values.   Whenever this is the case, the unavoidable perturbations, accidental or Soviet-fomented, would result in little permanent or cumulative damage.

In general, it would be wise to recognize and apply the principle that the development of reciprocal trade relations must obviously accommodate the existing regimes and power structures of the potential partners. It should not be carried, however, to the point where the stability of such relations depends uniquely or essentially on particular individuals being in positions of power and economic structures prevailing at a given time. Granted that it is more difficult to identify and to implement the gradients which transcend the idiosyncrasies of a particular regime, a few contemporary examples show that this is possible and rewarding.[20]

Perhaps the greatest difficulty in implementing the systematic development of the global gradient structure is the lack of motivation for devoting fairly large capital resources to internationally oriented ventures with relatively long-term payback periods. The stability of the structure depends on the continual replenishment of what could be aptly described as the "generalized stock-in-trade" of the West, the combination of knowledge, skills, products, services and capital that form the basis of attractive trade gradients as seen by the LDC's. The efficiency of the replenishment process may vary between rather broad limits, but there is no doubt about the magnitude of the effort required. The appropriate rates of the necessary investments are difficult to assess quantitatively, but a fair guess would indicate that they would involve a significant fraction of the rate of capital accumulation in the West, even if applied purposefully and efficiently. Over the long term, the cumulative investment of the West should exceed the aggregate investments by the LDC's, after taking into account the differences in efficiencies and initial conditions.[21]

---

[20] The ostensibly socialist and Soviet-leaning Iraq receives its civilian air transportation equipment, infrastructural support and operation support personnel from the U.S. The Angolan "revolutionary" regime, even with the highly visible presence of Cuban military, has retained and expanded its relations with Western oil companies to pursue off-shore exploration. The food exports to India do not seem to be affected by the ups and downs of the political relationships with the U.S.

[21] Literacy and working habits of the population, industrial base, infrastructure, etc.

It is not easy to channel capital investments toward long-term objectives even for domestic purposes. It may be most difficult to do so in support of international policy objectives especially when the stability of the global economic structure is still perceived by many as being threatened. Much can be done by the governments by clearly defining and publicizing their policies vis-a-vis the LDC's; by offering the same tax advantages and guarantees against expropriation as extended to other foreign ventures; and mostly by supporting those domestically oriented investments [22] which offer potential contributions to the creation and maintenance of international trade gradients. The interaction of these concerted measures may hold the key to the solution.

As the nature and the dimensions of the problem are brought into better focus, the traditional controversy of private vs. public initiative, competency, responsibility, and authority is bound to be raised. The arguments are very similar to those related to domestic issues; private corporations have the talent and many of the immediately applicable resources to tackle the problems of gradient development; they are doing it as a matter of course every day within their own sphere of interest and competency. What they lack, in general, is the long-term motivation which can not easily be conveyed to, and shared by, the stockholders, and seldom approved by professional managers constrained to show year-to-year "bottom line" achievements. Government agencies can in some instances develop the expertise and the long-term viewpoint required to guide broad-gauged and long-term efforts, but they are, both in the industrialized and the developing nations, affected by political constraints and shackled by the proverbial bureaucratic inertia. As for the U.S., there are pockets of expertise within the State, Treasury, Commerce, and Defense Departments and probably others, but nowhere (to my knowledge) is present the resolve or the authority to attack the problem of global economic framework development *in toto*. The National Security Council staff has the charter and the implied authority, but its interests are in general preempted by the

[22] Appendix—Note N.

relatively near-term crisis-dominated military aspects of international affairs.   Paradoxically, while we have a Domestic Council and a Council of Economic Advisors, there seems to be no agency or staff to coordinate these two with the internationally oriented concerns of other agencies. [23] If rapid progress is to be made in the direction advocated here, this deficiency should be urgently corrected.   Once the Western governments have clearly defined their individual and collective policy objectives, the requisite talent, motivation and expertise may be much more readily mobilized.   At that point, the normal division of responsibilities would keep the public authorities in the role of goal setting, monitoring and support, leaving private enterprise to carry out initiatives, and to bear the responsibilities and the ordinary risks in exchange for potential profits.   The remarkable vigor and flexibility shown by the multinational and transnational corporations demonstrate that private enterprise can flourish even in periods of rapid political changes; their expertise should be enlisted to build and to strengthen the global economic framework even when doing so transcends their immediate corporate interests.

Still as part of the considerations bearing on implementation, we must re-examine the currently uncontrolled process of "technology leakage."   This term applies to the unintended transfer of technical processes, skills, and more generally, the advanced knowledge pertinent to the means of production, without receiving appropriate value in exchange.   Our behavior probably originated at the time when the U.S. was a net importer of technology; but that period was essentially over by World War I.   It was thought at that time that the protection of trade secrets was fully achieved by the existing patent laws and by the industrial security enforced in their own self-interest by private corporations.   Today, unfortunately, none of these offer adequate safeguards; patent protection is often inadequate in a period of fast growth where the continuing investment rates, rather than the initial concepts, determine the future

---

[23]   The analysts of international development agencies, such as the World Bank, the IMF, or AID are probably the closest to representing the type of talent and expertise that would be required.

domination of a given industrial sector. As to industrial security, the mobility of top technical personnel, including specialists and managers, essentially frustrates both legal and practical safeguards.[24] Our generous attitude toward foreign students may in some cases attract their gratitude and sympathy, but quite often results in the unintended, or even explicitly subsidized, transfer of valuable knowhow.[47]

Historically, it can be shown that whenever a group allowed its "tool kit" (the primitive description of technology) to be transferred, without the "donor" society having preserved a much higher rate of technical advancement, it has invariably promoted the growth of eventually deadly competitors. It is not easy to determine whether continuing large-scale transfer of technology is in our long-term interest, but we may raise a valid question here in view of a recent lagging rate of R&D investment. Our major industrial trading partners (Japan and Europe) are far more cautious and restrictive in these matters. A new comprehensive U.S. policy in regard to technology management, involving the rate of development and the international transfers beyond the purely military-strategic considerations should be developed and implemented.

Mention should be made of two specific aspects of technology often discussed in the context of future relationships between the industrialized and the less-developed nations.

Appropriate technology refers to offering to the LDC's what they really need (as contrasted to what they want or they think they need); by definition appropriate technology will be directly usable by the recipients at the stage of their then current development. While it is certainly true that in some cases disparities have existed between the "high technology exports" offered by the West and the urgent, dire needs of the importing nations, such occurrences are normal at the initial stages of newly developed trade relationships.

----

[24] Note how the most elaborate precautions have failed to prevent Japan from acquiring all the essentials of integrated microelectronic circuitry; and how the activity of a few individuals, bordering or transgressing the threshold of criminality, has resulted in the effective transfer of substantial pharmaceutical trade secrets to Italy.

But when the expression "appropriate technology" is used in a derogatory and limiting sense to indicate that the West knows better than their LDC partners what is good for them, then the whole concept acquires a connotation which is increasingly resented and therefore inappropriate to the end in view.   In the very near future, the LDC's and the West should be able to determine together what technologies are appropriate by the simple mechanism of the marketplace.   To the extent that any new technology is required to satisfy a specific need, it will be by definition appropriate if the special requirements of the LDC's are properly taken into account.

Interactive technology refers to projects, equipment, and industries which, in addition to delivering the proposed product or service, also ensure long-term cooperation between the supplier and the purchaser.   In the most primitive embodiment, this concept would install manufacturing plants, for instance, in the host country, but would retain the supply of some critical element in the hand of the Western supplier. Soft beverage concentrates, sophisticated control equipment, critical high-technology components are representative examples.   In other cases, the supplier reserves a portion (preferably low volume, high technology and irreplaceable) of the production process to ensure continuous long-lasting interaction.   For military equipment, the maintenance and spare parts may be the basis of future interaction, with the concomitant political constraints. If the substance involves products entering the competitive world markets, then the continuous stream of innovations should constitute a most effective interaction mechanism.   While the objective of continuing, mutually beneficial interactions in certainly desirable, the concept of interactive technology will defeat its purpose if it involves the element of constraint.   The currently less-developed nations, if their basic interests are slighted by the interaction, will soon develop the work-around mechanisms aimed at eliminating any possible compulsion.

* * *

This is the end of another chapter, where the ultimate goals are clear and probably unimpeachable, but where the multitude of arguments makes it easy to lose the thread. A short recapitulation will be, I am sure, quite welcome.

A global economic framework based on multilateral, constantly rejuvenated, economic and cultural gradients is being advocated to complement that which is in existence now and which does not offer the desirable characteristics of stability and vigor.

Ideally, the desirable gradients are bilateral or multilateral, and include a significant fraction of services as contrasted to merely products and commodities. They must be consciously aimed at the long-term objectives of the participants. While government initiative and guidance are essential, private corporations and semipublic agencies can and should play important roles in the development and implementation of specific concepts.

The U.S. should identify and support the governmental focal point, possibly within NSC, where this kind of initiative is coordinated and nurtured, in concert with its domestic counterparts and those among our prospective foreign partners in this endeavor.

Increased understanding, of course, is always desirable, especially in undertakings of such magnitude. But the key to success in this particular case is whether or not the group of Western nations can, in the aggregate, generate the resources (skills and capital) necessary to replenish constantly the innovations which are the source of true economic gradients. Failure to so (which has been the case in many recent instances) will result in the substantial weakening, and the eventual disappearance of valid trade mechanisms. Such an occurrence would deprive the West of its traditional sources of wealth and the LDC's of their only reasonable chance for chaos-free, peaceful development interdependent with the West. This would be most unfortunate, especially in the next decade or so when all the fundamental factors promoting their development clearly point to the West as the sole nonthreatening source for knowhow and capital as well as practically an insatiable market for their products.

*

I have attempted to show that even when the differential rates of investment are preserved, a number of practical impediments remain in the path of the rapid global economic framework development. Some of them have to do with our understanding, some others are minor matters of organization or "modalities"; many of them are associated with distortions rooted in past circumstances and attitudes. The objective is so far reaching and so important that strong, well-coordinated efforts, involving the Western as well as the less-developed countries, are amply justified.

The ultimate measure of success, in regard to the LDC's, is the time span that will elapse before their societies develop the wealth, the sophistication, and the political system that, taken together, would prompt them to join spontaneously the Western commonwealth. The exact form of this adhesion is not important; in point of fact, the commonwealth may not even formally exist at that time. What is essential is that at that point, the population of the candidate nation essentially should share (about as spontaneously and enthusiastically as any nation-size group ever expresses a collective view) the generally accepted characteristic values of the Western political systems. At that time, and in regard to that particular nation, the international "gradients" will have become undistinguishable from those which drive the domestic economies of any Western nation. Let us hope that success, more often than failure, will be the ultimate outcome, but let us also fully realize that even with the most complete understanding, the best of possible motivations, and a large dose of good luck, the process will take a long time, perhaps decades. It would be naive to hope that the international economic system will be allowed to operate without perturbation for these decades of evolution. Whether and how to protect the system against violent perturbations is the even more difficult subject of the next chapter.

* * *

CHAPTER XIII

Military and Coercive Power

FOURTH PREMISE: THE AVAILABILITY OF EFFEC-
TIVELY USABLE MILITARY POWER TO THE GROUP OF
WESTERN NATIONS IS AN ABSOLUTE PREREQUISITE TO
THE ESTABLISHMENT AND PRESERVATION OF STABLE
POLITICAL AND ECONOMIC STRUCTURES COMPATIBLE
WITH THEIR COMMON IDEOLOGY, AT THE NATIONAL
AND GLOBAL LEVELS.

Very few people would take issue with this premise, except
those who are ready to live under any political system
rather than to face the horrors of future wars. Others will
object on the (mistaken) grounds that military preparedness
unavoidably leads to nuclear war, which is likely to bring
about the end of civilization "as we know it" or even the
destruction of life on this planet. What follows is not
addressed to those who hold these views. While moral and
religious objections deserve our respect, we should keep in
mind that they can be freely professed in our society only
because many of us are prepared to protect individual
freedom by all possible means—including force if that be
necessary.

The true message of our premise is conveyed by the specific
meaning attached to some of the key words. *Effectively
usable* military power should be credible in future conflict
situations and supportive of the policy objectives in view,
even under largely unknown future battle conditions.
Military power is an *absolute prerequisite* in the sense that
it should not be traded against other socially desirable
objectives; if the integrity of our political processes is not

effectively protected against perversion or destruction by violent means, none of the other social goals will be even debated, let alone achieved. *Compatibility* with Western ideology shared in common is stressed as one of the essential characteristics of military power advocated here; it must not be used for the purpose of imposing Western political views on any group or nation unwilling to accept them spontaneously according to the very standards we wish to preserve; furthermore it should not be used for aggressive or exploitative economic purposes, but should be effective in preventing pursuits of this type initiated to our detriment by our competitors.

Before discussing the specifics, a few preliminary remarks might be useful. There are still some, otherwise well-informed, people who are prone to question the necessity, or at least the urgency, of changing our approach to preserving or restoring our military power. Let us all be reminded that a modern society, such as the U.S., is remarkably fragile as a system[1] in spite of its imposing physical appearances and the exquisite refinement of its artifacts. Civilizations and empires have ceased to exist in the past for both political and ecological causes, but except in rare instances, their demise has taken place over several generations.[2] In modern times, however, major shifts in political fortunes and alignments of even powerful nations may occur unpredictably with catastrophic swiftness: Imperial Russia has crumbled and its successor has reached superpower status in a matter of five decades; the rise and fall of the Nazi Third Reich took place in 12 years; China was conquered by its current rulers in less than fifteen years centered on World War II;

---

[1]   It is easy to find many references to the fragility of the environment and to the vulnerability of "endangered (animal) species." It is undoubtedly true that uncounted millions of species come into being and disappear with amazing swiftness as measured on the geological time scale, but taken as a whole, organic life is one of the most resilient components of the Earth's crust.[48] By way of striking contrast, one may find references to the vulnerability of the human species but practically none to the fragility of industrialized, and more particularly urban, civilizations.

[2]   The Roman Empire was formally "destroyed" in the 5th century A.D., but its formal successor was very much alive almost a thousand years later, until the fall of Constantinople in 1453.

the British, French, Dutch, and Belgian colonial empires disappeared essentially in 20 years—the coalition of "petro-powers," which appears to carry some influence in world matters, is barely 10 years old. Indeed, major changes occur in the political world at a rate which will appear bedazzling to future historians, assuming that they will still be around and interested in what happened to our century. Be that as it may, such changes may well affect the Western nations, and more specifically the United States, in a major and possibly catastrophic manner within a matter of a decade, and certainly within the lifetime of most of us.[3] It is not too early for strong and purposeful action; there are only too many who try to convince us, as well as themselves, that it is already too late.

There is absolutely nothing in history, long past or recent, which suggests that anything less than military power decides the ultimate destiny of political causes, no matter how meritorious. One may even observe with a touch of cynicism that merit, value, and virtue are in fact determined by the outcome on the battlefield.[4] Even such venerable institutions as the British House of Parliament and the parliamentary system of government trace their

---

3 Premonitory signals are clearly perceptible. Just as the generation of the American Founding Fathers watched with dismay and apprehension the courtiers of George III impose arbitrary taxes and regulate the details of their day-to-day existence, we observe with fascination the semi-annual pageant of the OPEC Council meetings. Here the disparately garbed high-ranking delegates (often the polished products of Harvard or the London School of Economics) hold forth in front of microphones and explain in a variety of foreign accents their lofty principles and motivations. Between lavishly served meals and sumptuous entertainments, they quietly decide about our Sunday driving habits, our family budgets, our balance of trade, and many other important or trivial facets of our daily lives. There is nothing more galling or more abjectly telling about our relative political strength than the two-inch headlines on the morrow of such conferences announcing triumphantly to our relief that the "moderates" among the Council members (after a bitter struggle, no doubt) have prevailed upon their colleagues to limit the rise in their prices to only 28% this quarter...

4 In the past few decades we have had ample opportunity to watch the subtle gradations in journalistic terminology: bandits become terrorists when they fight the wrong side, but sublimate into guerrillas, patriotic fronts, freedom fighters when they are anywhere close to military victory.

origin back to the Civil War (1640-1649). As to the U.S., one shudders at contemplating the fate of the Founding Fathers had the War of Independence ended in the defeat of Washington's armies. More recent events lead to equally firm conclusions: All around the world, in Russia, India, Israel, Indonesia, Indo-China, and Black Africa, and in just about every other instance, the current political power was acquired and consolidated by means of violent struggle, war, civil war, revolution, "coup"—the exact technical designation is immaterial; what counts is the successful use of brutal, naked force. The argument that war is obsolete in the nuclear age finds its overwhelming refutation here; except for the U.S.S.R. all the above examples are drawn from the brief time span elapsed since 1945.

This chapter briefly discusses the background and the rationale for changes in Western military power, offers a few pointers in the direction of effectiveness and affordability, and concludes (convincingly, I hope) that not only is military power an absolute requirement, it is also perfectly within the realm of practical possibility if the West will just stop apologizing for defending itself and start applying to the task the ingenuity and, most importantly, the tenacious resolve which was historically the true major source of its greatness.

*

The primary driving force behind the urgency of reassessing the Western approach to military preparedness is the relentless Soviet thrust in the direction of global supremacy.

The Western world is in direct military competition with the "Eastern bloc" [5] composed of the Soviet Union and its satellites. In spite of its smaller resource and population base, in spite of its much lower GNP and aggregate growth rates, the performance of the Eastern bloc in terms of military capability development in the past two decades has

---

[5]   This term is used merely as a convenient shorthand designation. It is not aimed at conveying the sense of cohesion or homogeneity.

been nothing short of impressive. Supported by the satellites' relatively advanced industrial base and by Western technology obtained as one of the benefits of the "detente climate,"[6] reinforced by the highly respected role played by the senior military in their system of government,[7] the Soviets have achieved clear superiority, or at least acknowledged parity, with respect to the West in all important areas of military capability.

Their strategic nuclear strike forces surpass the combined U.S., British, and French inventory by comfortable margins, whether measured by delivery vehicles, aggregate weapon yield, or target kill capability; very soon, even with the SALT II limitations, the balance in the number of individual warheads will also tilt in their favor.

The Soviets have currently deployed long-range theater nuclear weapons, capable of covering all of Western Europe; this is independent of their relatively new "nonstrategic" bomber which, depending on its basing, operational employment, and ordnance, can be used in the strategic nuclear or theater nuclear role, or again as a major long-range threat against the Allied naval surface forces.   The Soviet and Warsaw Pact general-purpose conventional forces are being advertised and acknowledged as being far superior in numbers and comparing quite respectably in quality to those available to NATO; this balance may be even far more favorable to the Warsaw Pact than the comparison of "conventional" assets might indicate.[8] While the immediate

6   Vigorously advertised by both East and West, with only the latter's military investments being restrained by its spirit and intent.

7   Note the longevity of the top Soviet commanders. A quarter of a century tenure is nothing unusual among the heads of military service departments; this may impose a measure of doctrinal rigidity, but certainly helps continuity in the pursuit of longer term objectives.

8   There is substantial and reliable evidence on hand to indicate that the Warsaw Pact forces have the equipment and training to fight in battle environments created by nuclear, chemical, and possibly biological warfare; the Soviet forces are well supplied with battlefield nuclear weapons as well. In the central NATO front, geography offers further advantages to the Soviet dominated Eastern bloc. While their initiatives benefit from a broad front with multiple and redundant lines of communications, a significant portion of NATO strength relies on U.S. based airlift and sealift operations designed to reinforce the relatively weak prepositioned assets.

Western concern is focused on the NATO-Warsaw Pact balance, it should be obvious from the global Western viewpoint that the Soviet equipment and the combat readiness currently aimed at the European theater might eventually play a major role in other theaters as well.

The Soviet thrust toward naval power at the global scale is another example of sustained, purposeful investment in support of long-term military objectives. In addition to a growing number of relatively modern cruisers, destroyers, and small missile-capable surface ships, the Soviet Navy will have by the mid-80's at its disposal aircraft and helicopter carriers, clearly aimed at sea control and credible overseas force projection. The submarine technology acquired during and following World War II, coupled with the advent of nuclear propulsion plants and with the strategic importance of sea-based ballistic and cruise missiles, has given the Soviets numerical preponderance in all aspects of submarine warfare. Their antisubmarine capability is less well known (mostly owing to the multicompartmental military security filtering the available intelligence information) but that it has not been neglected in their overall drive for naval supremacy may be inferred from numerous relevant references in the Soviet military literature. The deliberately publicized "OKEAN" exercises also reveal growing worldwide, integrated naval command and control capability.

One could pursue this recital of specific areas where the Soviet thrust has already altered the military balance or is in the process of doing so, but the basic point has already been made. The Western world is not at leisure to continue its present military course; doing so would mean sooner or later to be confronted with only two alternatives: In any crisis or escalating military conflict, to face, with inadequate preparation, directly or indirectly, the highly credible and growing Soviet military power; or, well before the crisis has even reached the overt military combat stage, tamely and abjectly submit to whatever our opponents choose to request at that particular time. From the particular U.S. vantage point, while the current dedication of its Allies to the Western democratic ideals might still be strong, support and cooperation or desire for sacrifice will evaporate as soon as the aggregate military (and therefore political) leverage in crisis situations will have been proved repeatedly deficient or nonexistent.   Both alternatives, I submit, are disastrous; so this discussion is pursued on the basis that major modifications in our current approach to ensure (or to develop) continuing, adequate military posture are urgently required, and perhaps long overdue.

*

Recent changes in the political, military, economic, and technology environment suggest the need to reexamine the mechanism whereby military capability requirements are established and may alter some of the related, strongly held, current premises.

A bewildering array of major new military and technical developments are likely to change the nature of warfare almost beyond recognition with respect to World War II and even the Korean War. Just to set the stage, a brief review of these, without any particular order or priority, might be useful at this point.

Both superpowers have space-based reconnaissance and surveillance capabilities[9] augmented by high-altitude airborne components.   Soon the same capabilities will be directly or indirectly available to second-rank military powers.   Coupled with the long-range delivery of relatively powerful weapons, this capability threatens all major assets of the combatants which are "visible," i.e., accessible to enemy acquisition. While this was often the situation in the past in the vicinity of the battle zone, it now covers all the directly engaged assets, the second and rear echelons including the logistic system having its roots all the way

---

[9]   Without attempting the rigor of military taxonomy, these terms are used in reference to sensory systems capable of acquiring information about the nature, location, and activity of enemy forces, in particular those on the ground and sea surface, those aloft in and above the airspace, and those submerged in the oceans. Electronic intelligence is specifically included, but not the intelligence obtained through the classical human "intelligence" channels.

back at the landing ports or even overseas. Put otherwise, except for self-imposed restraints, any military engagements may easily and rapidly acquire global dimensions. Owing to its particular geographic and political situation, the U.S. is particularly vulnerable in this respect, but as the U.S.S.R. reaches for objectives farther away from its direct land borders, it will become similarly exposed.

Owing to their vital role in reconnaissance/surveillance and also to their important contributions to communications and navigation, satellites are likely to become attractive and time-urgent military targets. While this area may be operationally beyond our time horizon,[10] the research, development, procurement, and training related to "space war," including space, ground, and airborne elements, will attract the growing interest, and absorb an increasing portion of resources, of the leading military establishments.

While dwelling in the realm of the exotic, the possible role of new type of "weapons" should be mentioned. The age-old technique of throwing lethal objects against each other has certainly evolved from the hand-held stones, branches, and nuts of the proto-hominids all the way to the precision-guided missilery of the late 20th century, but the principle has remained essentially the same. The advent of focused energy beam weapons, if feasible and practical, may introduce a difference in kind as well as in magnitude, which could quite substantially detract from the popularity currently enjoyed by the guided missiles.

At the other extreme of the conflict spectrum, we should recall the points emphasized earlier in this study. A number of military conflicts, perhaps intrinsically secondary or even trivial but attracting attention because of the associated escalation potential, will involve relatively primitive combat environment. Guerrilla, urban terrorism and their variants will be with us for quite awhile; once motivated or coerced into purposeful, sustained military or paramilitary action, relatively primitive groups can inflict damage on their more sophisticated opponents. It is possible that the

[10] Arbitrarily chosen as A.D. 2005, i.e., 25 years from the time of writing.

growing prevalence of such nonconventional combat techniques will increase the pressure for nonlethal, but highly pervasive chemical and biological weapons—moral revulsion among the advanced nations notwithstanding.[11]

Between the esoteric domain of space wars and the grimy, unglamorous business of guerrilla warfare lies the broad spectrum of "intermediate scale" engagements, where both sides engage their first or second line combat equipment and where the possible employment of nuclear weapons must be seriously considered. On land, at sea, and in the air, this type of warfare has been subliminally pictured as massive encounters with brute-force-driven communist "hordes" blundering into the exquisitely sophisticated high-technology strength of the West—but then, the scenario stops right there. We know by now that the technology levels of the two sides will be at least comparable; as to the reliance on high technology, it still remains to be proven in protracted major engagements. Specifically, the heavy concentration on precision-guided missiles requires a rich, prompt, and highly reliable information flow which may not be available in real combat situations.[49] The importance of military information is now so well recognized that the associated command and control systems have become combatants on their own right and subject to catastrophic failures even while retaining their physical integrity. A related problem is the actual availability, flexibility, and performance of highly sophisticated electronics under combat conditions, especially when maintained and operated by military personnel with less than adequate educational background and training.[12]

---

[11] To be noted that the Soviet military are rumored to have reached the same conclusions in Afghanistan. The reaction of their public opinion has yet to be observed, but then neither was the uproar of the world-wide leftist establishment particularly deafening.

[12] In this and related connections one should recall that we may refer to ad-hoc Allied personnel not sharing the Western industrialized societies' skills and mental attitudes. This is not to say that many of them are not highly intelligent, motivated, or even civilized according to their own cultural standards; but that is a far cry from proficiency in the operation and maintenance of sophisticated equipment.

In addition to the technical and conceptual aspects of future wars, the structural characteristics of the opposing societies are also apt to contribute quite decisively to the nature of military preparedness. Thus, we have seen that the traditional strategy of the West consisted in engaging almost to the last minute prior to overt hostilities in business-as-usual conduct of their daily affairs and even in trading with the potential enemy. Once the crisis reaches the stage of military action, we expect to "mobilize" and then, ever so slowly, to gear our industrial resources up to war footing. In all probability, this type of leisurely approach will not be given the opportunity to succeed. We must rather envision rapid crisis evolutions coming to a military climax in a matter of days or even hours; thus, both the time rates of change and the possible interplay of simultaneous events in far distant theaters renders instant readiness and the ability to rapidly change deployment postures mandatory. The current U.S. and Allied approach, consisting of protracted requirements definition, design and development phases, followed by minimal procurement of operational equipment lends itself poorly to demands for this type of readiness and flexibility.

There is still another important area where the characteristics of openness of the Western societies create a potentially disastrous dissymmetry in favor of its opponents. Institutional restraints, such as arms limitation treaties, may appear to offer some hope to reduce the psychological pressures of an unlimited arms race and also possibly the level of military investments. For these reasons, arms control negotiations are, in general, popular with the electorate and continue, often beyond the point where the potential dangers outweigh the possible benefits. In a world where the penalty for error might well be the choice between surrender or defeat in a nuclear war, reliance on such institutional restraints, without efficient insurance-type programs as a protection against treaty violations, is a risky course to take, but this is exactly what we are doing at the present time.[13] Here again, the open character of our

---

[13]   When the Soviets appear to violate any provision of the SALT treaty, the U.S. reaction is either to explain it away by publicly supporting the view

society, coupled with our inability to respond fast to blatant challenges, creates an element of weakness. We have a tendency to respect treaties and any internal suggestion of "cheating" is usually received with considerable disfavor at senior policy-making levels. By way of contrast, the Soviets are in no way held to this kind of restraint. Violations are not only possible, they are also likely to be officially encouraged, or at least tolerated.[14]

A final point must be made on the subject of the motivations of the individual combatants. Those responsible for the Western military preparedness will have to contend with the increasingly manifest circumstance that young adults accustomed to reasonably high standard of living simply do not relish protracted military service in times of peace, and even less the hardships of active combat conditions. This is in strong contrast with the far more regimented and indoctrinated manpower of the "socialist" regimes. The contrast is even sharper with respect to the teeming masses of Third and Fourth World teenagers who, having literally no possessions other than their feral hunger and their biological pressure to procreate, will risk their lives in combat on terms which are clearly irrational to Western understanding.[15] In the past, especially for relatively remote and small engagements, even the West could depend on heroic and romanticized specimens of martial virtue such as the Foreign Legion, Green Berets, Commandos, and other elite troops, but this unfortunately is no longer so. We can no longer expect to compete on a man-to-man, or better,

13   (Footnote continued)
    that the violation, in fact, has never occurred or to condone it tacitly.
    There is very little else to choose from, as a matter of practical possibility.

14   A glaring example is the history of negotiations on the Comprehensive
    Test Ban on nuclear weapons. Even while negotiating, we already sedu-
    lously cut the budgets of the defense laboratories and encouraged the dis-
    banding of almost irreplaceable teams of experienced scientists/designers
    engaged in weapon development. The Soviets have absolutely no problem
    in maintaining their specialized laboratories under any overt or covert
    designation; they keep on preserving their scientist/development teams and
    thus derive maximum benefit from the "peaceful" nuclear explosions.

15   Professor R. Fox, personal communication.

person-to-person basis, unless our combined motivation, organization, and equipment are deliberately designed for this purpose.

***

Some of the major military mission capabilities, long taken for granted as the guarantors of basic Western foreign policies, are not viable in the new environment.[16]

Central Strategic Nuclear Balance and Deterrence

The strategic nuclear deterrence was for several decades accepted as the key to the protection of the U.S. homeland and to its ability to control ("manage") escalation in local, regional, and theater-wide contexts. Its credibility is contingent upon the Western preparedness to actually fight and prevail in a central nuclear war, should the deterrence fail. The deterrent stability depends on the several contributory offense weapons systems and the associated command and control structure being immune to surprise attacks ("preemption") conceivably seen by the Soviets as offering substantial incentives to strike first.

Knowledgeable analysts have reached the conclusion that the central nuclear war assets of the United States and of its Allies are, in their present state, vulnerable to pre-emptive surprise attack, that the surviving weapon and command and control assets also lack sufficient endurance[17] to credibly engage in a protracted nuclear exchange with any serious chance of prevailing against the residual Soviet forces. The current ability to limit damage to the U.S.

---

[16] This and the following sections are mostly written with the United States as the primary focus and vantage point. In general, the observations are thought to be applicable with only minor modifications to most Western military powers, unless specifically stated otherwise.

[17] This observation is also valid for submarines beyond a few weeks at most following the outbreak of hostilities.

population and industrial base in a central nuclear war is questionable; in the opinion of many it is practically nonexistent. The comparison of *resiliency* of the Western versus Soviet societies, in the sense defined earlier,[18] is, in general, thought to favor the Soviets.

As the Soviets acquire competency in submarine warfare, and as the ability of the U.S. and Allied navies to effectively control the open seas is being brought into question, the survivability of ballistic missile carrying submarines must also be reexamined. Submarines in international waters are prone to covert attrition without requiring highly conspicuous and attributable nuclear weapons. The U.S. has no credible capability to protect, or to react, against this threat, except overt warfare—a most unattractive option in all circumstances.

The arms control limited environment, as it has existed since June 1972, has failed to inhibit the progressive deterioration of strategic balance which theretofore was in favor of the U.S. It is expected that, unless vigorous unilateral action is taken, such deterioration will further continue.

Theater Strategic Balance and Deterrence

If the credibility and stability of deterrence favoring the U.S. and its Allies in the central nuclear war context is questionable, the Allied nuclear war posture in the theater context is even more so. The vulnerability of the NATO "strategic" weapon bases, as currently operational or planned, offers substantial incentives to preemptive surprise attacks; furthermore, the existing and currently planned weapon systems have insufficient target coverage and reaction times too long to serve as effective counter to the now deployed Warsaw Pact attack capability.

[18] Chapter VIII.

The deficiencies in military mission capability introduce political issues which further bring into question the viability of the Allied theater strategic posture. Thus, the vulnerability to preemption raises the well-justified fear of "collateral destruction" of their cities among the Allied host population—a factor not precisely calculated to enhance the political acceptability of such weapons. The U.S. will have continuing difficulty in convincing its NATO partners that the combination of its own commitment to the Alliance and the military effectiveness of theater nuclear weapons can counterbalance the risk of nuclear devastation.

Force Projection and Blockade

More generally, the ability to wage wars in overseas theaters requires for the U.S. to "project" its forces to remote battle areas where previously established military infrastructure is not available. At the same time, the access of hostile forces to the same battle area must be denied through "interdiction" (blockade) of all air, land, and sea routes. Conversely, the enemy forces will attempt to keep our forces and their logistics support out of the combat zone; moreover, when supported by Soviet long-range acquisition and weapon delivery capability, or by Soviet-supplied submarines, they will attack the friendly forces in transit. Here again, the survivability and the effectiveness of a long logistic chain, dependent on high-value and targetable vehicles and vulnerable terminal points (port cities, cargo handling facilities, fuel supply lines), render overseas operations in the face of determined opposition most hazardous.

It is conceptually possible to provide survivability to the force projection assets and to its logistic chain. This involves the dispersal, the passive protection, and the active antiair/antimissile defense of the fixed facilities; ships and aircraft will have to rely on their mobility, camouflage/ deception, but mostly active defense by naval forces and combat aircraft.[19] Such capability maintained in a state of

19   These, of course, require their own operating and support bases, which also need protection.

readiness, including frequent exercise, over years of peace-time or subcrisis-level situations is hardly an attractive proposition from the cost viewpoint.

But even if we have elected to invest in a permanent well-planned and well-exercised infrastructure, which renders our posture advantageous in a certain sector of a given theater, the opponents (possessing the initiative *by definition*) can stimulate in a matter of a few years or even months highly threatening crisis situations in other theaters which would force the West to duplicate and maintain simultaneously such expensive readiness in several places. Intertheater mobility coupled with effectiveness and survivability are not easy to achieve under any circumstance, but just about impossible under stringent cost constraints.

The Role of Manned Aircraft

The aircraft's role should be reexamined in the light of new combat environment. Its desirability as a strategic or tactical combat system or as a transportation element is not in question but its continuing dependency on airfields and soft logistics support in an environment where the enemy has at its command long-range target acquisition and precision weapon delivery render its overall combat endurance questionable. The use of aircraft carriers loses much of its credibility when the carriers themselves fight for survival (let alone for operational capability) as they are almost certain to do in any large-scale encounter possibly involving nuclear threat.

Aircraft also must fight a much changed defense environment if operating anywhere near the enemy defense perimeter. In an era of rapid and precise information processing, long-range acquisition of aircraft (even those which attempt to fly in the nap of the earth) is possible by means of advanced airborne sensors; the ground defenses also have much increased their competency by resorting to mobility, distribution, and netting. A broad panoply of competent surface-to-air and air-to-air missiles is at hand to challenge even the most sophisticated aircraft at a fraction of its cost.

> "...Air engagements involving first-line forces of the technically advanced nations are seen as encounters between a large number of offense and defense elements, many internetted sensors, and processing and decision-making nodes. These will include aircraft of several types, most of them strenuously attempting to keep out of the reach of the enemy's weapons. The burden of survival and of success will be placed on an *electronically integrated multielement structure*, as contrasted to individual aircrafts mostly dependent on their aerodynamic prowess, aided by the skill and the heroism of human pilots..."[49]

It is far from clear that the currently planned tactical combat aircraft inventory is responsive to the challenges, or capable of exploiting the opportunities, inherent in this characterization.

Aircraft Carriers[20]

"Battle groups" centered around nuclear-powered aircraft carriers are the core of the currently planned U.S. naval power. Carriers are vital for both sea control and force projection missions; they also contribute the major share of the Navy's theater nuclear forces. The procurement of nuclear carriers is a lengthy and costly process but no practical alternative appears to be present.

New technology developments pose generically different new threats to aircraft carriers: (1) space and high-altitude aircraft can identify and localize major surface ships in near-real time within accuracies consistent with effective long-range weapon delivery; (2) targeting information can be transmitted to, and used by, a variety of long-reach weapons, including cruise missiles with nuclear or other

[20] Added in proof.

types of lethal warheads. Land, air, and submarine based versions of these weapons can be conceivably synchronized to create multi-directional simultaneous threats to the carrier. Supersonic and terminally guided attack weapons, launched beyond the acquisition range of the carriers and of their auxiliaries are within the bounds of realistic threat projections.

The issues seen as associated with our carrier programs are: (I) the conceptual alternatives at hand (to nuclear powered large aircraft carrier) to offer improved survivability in the projected threat environment, without sacrificing the carriers' offensive capability; and (2) the adequacy of current development programs aimed at protecting the carrier battle groups against synchronized multiple high-penetration capability attacks, in view of the timing and magnitude of the threat projections. The ability of carriers to survive and to operate in a protracted nuclear battle environment is a matter of considerable concern, unless the U.S. central (strategic) nuclear deterrence remains (or perhaps becomes again) unchallengeable.

Balanced Emphasis Among Nuclear and General Purpose Forces [21]

The need to reverse the erosion of our strategic deterrent capability is widely accepted as being urgent and imperative. The possession of capable and credible strategic nuclear forces will reduce the danger of deliberate nuclear attack against the U.S; it will also discourage Soviet-initiated escalation of lesser conflicts to the central nuclear war level. However, in the foreseeable future when the Soviets have in their possession powerful nuclear offense weapons and when the U.S. cities and industrial targets remain essentialy unprotected, strategic nuclear forces can not effectively deter lower level aggression. It is quite unlikely (and certainly not believed by potential aggressors) that the U.S. would resort to the use of nuclear weapons in response to any provocation except overt, major attack

[21] Added in proof.

against U.S. territory. Be that as it may, the use of strategic nuclear weapons should certainly not be the only option available to the U.S. National Command Authority. Thus, if we are to prevent a continuing succession of lower level hostile initiatives, all cumulatively reducing the U.S. influence, power and freedom of action, and if we are to protect our Allies against political and military coercion, effective forces capable of dealing with all distinct levels of conflict are necessary. We must be prepared to fight high-level conventional wars (including the possible use of chemical weapons) and we must also be in readiness to counter, or to initiate, theater-level nuclear warfare. It should be clearly perceived, however, that the effectiveness of our general-purpose forces can be fully negated if the enemy has superior theater nuclear forces; by the same token our theater nuclear forces can be inhibited if the Soviets are conceded strategic nuclear superiority. Strategic nuclear forces are thus seen as essential to deter central nuclear war and to enable our theater nuclear and our general purpose forces to accomplish their respective missions. The strategic nuclear forces are unsuitable to the task of deterring or fighting lower lower conflicts; we must therefore provide effective military capability-in-being at all levels. The issues of relative priority and emphasis among strategic nuclear, theater nuclear, and general purpose forces are fallacious. All three must be present and effective; together they can support the whole broad spectrum of our policy objectives, but none of them can do so by itself in isolation.

*

We have just examined a short list of intermingled political, military, technical developments; many of these point to areas of weakness and some to glaring deficiencies in our defense posture. With even the most cursory effort, the list could be considerably expanded and each heading would imply the same dreary conclusion: attempt to do more, increase the capital and operating budgets, seek salvation in better technology...and we already know that such increased commitment to military preparedness is not likely to be palatable to democratic nations on a sustained basis.

Certainly a significantly different approach should be taken in our military capability development process. Generic principles must be found which could conceivably render our investment in military strength more effective, more consonant with the nature of future major wars as now foreseen, and more compatible with the very nature of the societies which are the presumptive beneficiaries. A possible approach to this task is to be discussed presently.

* * *

The general principles used to guide our military posture developments must capitalize on the intrinsic Western strengths and exploit the adversary's weaknesses. Those explored here involve the diversity, the dynamics, and the information flow associated with weapon development and deployment, as well as the synergies between military and civilian oriented resource applications.

Several weighty tomes could be written on such a rich topic, but it is fortunately beyond the scope of this work (and the competency of this writer) to offer a complete and exhaustive treatment of the subject. This emphatically is not a blueprint for the desirable structure of our future military forces, nor is it a philosophical discussion on military operations or procurement procedures. It is what it is advertised to be, the outline of a few general principles that might be found helpful in formulating more detailed plans[22] for defining the desirable structures, operations, and procedures.

Again, a note of caution. This is a complex subject; many of the ideas discussed here are present in many other forms and under many other titles. Originality and uniqueness cannot be claimed for this particular organization, neither can it be said that the matters discussed under each heading are fully or even substantially independent. Individually,

---

[22] A companion volume[50] is planned to give the more specific, hardware-oriented discussions.

and in combination, though, they are seen as illuminating the several pertinent facets of the subject.

Multiple Complexions

Deterrence is a rare word: even nowadays, in year 35 of the nuclear age, it is used exactly with the meaning given by the dictionary. [23] Mostly fashionable in the mid-1960's, deterrence *by fear* was embodied in the concept of "unacceptable damage" to be inflicted upon the Soviets if they dared to misbehave, especially in the central nuclear war context. Deterrence *by anxiety* could be interpreted in military terms as the "exchange ratio" of the weapons engaged; in short it means "don't you start anything; you will be worse off for having tried." Deterrence *by doubt* is our essential concern here; if the enemy cannot predict, with any reasonable degree of confidence the outcome of his initiatives, he will in fact be deterred.

Most of the discussions center on deterrence of nuclear wars, but we should observe that it does operate really at any and all levels of crisis and warfare. Since the West is fully oriented toward the preservation, or at least the nonviolent, slow change of the *status quo,* the prime role of the military is to dissuade the enemy from starting or intensifying ("escalating") moves hostile to our side. The dynamics of escalation are controlled by known (but mercifully unverified) laws; they are exceedingly difficult and costly to apply, if the enemy is consistently given the advantages of initiative, surprise, and essentially complete information of our "order of battle" (i.e., the location and the nature of the assets we intend to engage in any particular encounter).

The problem is how to ensure effective deterrence at any level, in the presence of these adverse conditions. In the incipient phases, i.e., prior to overt hostilities (or to signifi-

---

[23] To discourage or keep a person from doing something, through fear, anxiety, doubt, etc. (Webster)

cant escalatory steps) we must effect deterrence by doubt; we must ensure that the enemy cannot predict the outcome of any of his hostile initiatives with any degree of confidence.[24] Should this type of deterrence fail, once military operations have started or escalated, we must resort to anxiety and fear: the enemy must be made to realize that (with our assets essentially intact and available) we *will* prevail in the subsequent engagements.

In the states of incipiency we must not only have the appropriate forces available to fight the battle with reasonable chances of success, we must also prevent the enemy from destroying these forces prior to the actual engagement.[25] In addition to making the forces intrinsically survivable (i.e., difficult to destroy), the most effective way to protect them is to make it difficult, or better impossible, for the enemy to target them effectively. This can be done,[26] by multiplying the number of *complexions*, i.e., the time-variable combinations and interactions among all possible basing modes and all pertinent deployment areas. Not only must we provide multiple complexions, they also must operate in a synergistic, beneficially interacting way, forming a number of combinations beyond the mere juxtaposition.

Beyond the stages of incipiency, once the engagement has started, we, of course, again resort to the advantages afforded by multiple complexions...but by emphasizing this point, we are simply rediscovering the essential need for tactical flexibility. Many types of weapons, multiple and dispersed units, the value of surprise, the synergistic use of forces, all these are old friends of the experienced military. The use of rapidly flexible tactics complicates the task of the enemy, degrades his tactical decisions, and thus increases the chances of victory for the friendly side. This

[24] The term "assured confusion" (applied to the enemy, of course) to designate a deterrent principle is attributed to Dr. R. Latter.

[25] Forces, in this context, include the weapons, the personnel, the command and control, and the logistics support systems.

[26] Appendix—Note T.

in fact is so well understood that, in our mission requirement definition process, it has tended to overshadow the
"deterrence by doubt" in the incipient phases.   In the
currently planned Western force structure, there is heavy
overemphasis on the fighting ability, and relatively
insufficient attention paid to the stable deterrence in the
incipient states, i.e., the survivability of the friendly
resources prior to actual engagement.[27]

What is new and different in this view is that it recognizes
that the term "battle zone," in general, has become
meaningless. When several escalatory phases are or can be
intermixed, the resources of a whole sector or even a whole
theater are brought to bear on the battle outcome. There is
no clearly defined FEBA; weapons of all types and ranges
are engaged. It is already obvious that there are no land
combats of any significance without close air support, air
superiority, and air interdiction type operations; very soon
medium- and long-range missiles will complement, and
eventually overshadow, the traditional artillery firepower.
In different contexts, naval combat has long ago
transcended the traditional ship-to-ship engagements and
soon the task-force-level encounters will be augmented by
the intervention of theater-level resources, i.e., long-range
combat aircraft and long-range cruise missiles. Both the
concepts of survivability in the incipient phases and that of
fighting ability during actual combat operations take on new
aspects and it is argued here that the multiple complexions
in our weapon deployments and characteristics are an
important prerequisite for effectiveness.

Whereas a number of ideas associated with multiple
complexions in warfare have been elaborated in the context
of central nuclear war,[51,52] we must apply the same
principles to lower level encounters in all theaters and all

---

[27]   In the central nuclear war context, the "fighting" ability has unfortunately
centered in the first few hours of combat, in keeping with the massive
retaliation and assured destruction concepts. The *enduring* capability of
our strategic nuclear forces has received relatively little attention, probably
on the (mistaken) assumption that fighting nuclear wars is unthinkable and
winning them is impossible. The Soviet military, according to their published doctrines, do not appear to share this view.

combat modes. Specifically, naval surface warfare, undersea warfare, space warfare must be reexamined from the viewpoint stated here. The proposed concepts for overseas force projection (including the currently emphasized Rapid Deployment Force) and for theater nuclear warfare deserve special and urgent attention. Not only the weapons and the troops, but all other assets, such as the command and control system and the logistics systems, can be targeted, and therefore must be supported by the same principles of multicomplexioned deployment so as to enhance survivability and fighting ability.

The principle of *multicomplexioned warfare* is recommended as an element of major advantage to the "nonaggressor" side of future conflict. Its systematic application to weapon deployment, command and control assets, and logistics support systems enhances stable deterrence in the incipient phases of conflict and escalation; it also strengthens our fighting ability subsequent to deterrence failure.[28]

In this conceptual context, we are helped by the Soviet doctrine and tendency of relatively massive investment and rather rigid doctrinal training. We are perhaps helped also by our industrial diversity and our (potential) ability to manage relatively complex operations. This assertion will have to be verified.

Dynamics of Posture Changes

Posture in the present context refers to the combination of material and intangible assets which can be applied to achieve our national security objectives. It includes military forces in readiness, the reserves, and the mobilization potential, the organization and training, as well as the morale and motivation of military and civilians alike. The

---

[28] If the reasoning given here is correct, if it is true that the deliberate introduction of multiple complexions offers major advantages in terms of survivability and fighting ability, the concept of multicomplexioned warfare will eventually spread to our adversaries. It is not too early to start thinking about our countermoves in that regard, but it is left to future revisions of this work.

availability and reliability of information, such as warning and intelligence, are, as always, vital components of our posture. At all levels of conflict, but more particularly in nuclear wars, the population and industrial resources are essential as well.

Posture changes result from unilateral initiatives taken by one of the adversaries, followed by a more or less specific response by his opponent. The purpose of the response being to deter further action by the initiator, it must be perceptible, appropriate, and timely. It must clearly convey to the "aggressor" that we are ready and have the resolve and the capability to defeat his purpose; it must just as clearly and convincingly convey the same message to Allies and neutrals whose attitudes may have bearing on the situation. The response in terms of our posture changes must be *appropriate*; i.e., strengthen our deterrent and fighting capability sufficiently to reverse the course of escalation, but not so overwhelmingly powerful or provocative as to trigger further irrational enemy actions. *Timeliness* in the development of the posture change is, of course, crucial; if our response is sluggish or long delayed, its effect will have no bearing on the outcome; if it is on the other hand too rapid, it may be perceived as an aggressive threat by the enemy, calling for further "response" and thereby causing the very escalation we wanted to avoid.

In the particular situation of the West, the peacetime cost of maintaining in a state of readiness military forces able to overcome *any* Soviet initiative is probably prohibitive, so our ability to maintain stable and reliable deterrence is based on the appropriate *dynamic characteristics* of the posture changes we may have to effect at the time when the Soviet move is being perceived. This is, by far, no idle concern; the Soviet Union has amply demonstrated in the recent past that they would not hesitate to introduce surprise-type initiatives into the delicately poised East-West balance at any time when they can sense a possible advantage. An incomplete list of past instances would comprise:

- Launching of SPUTNIK I (1957)

- Breaking the nuclear test moratorium, by a long-prepared massive series of atmospheric tests (1961)

- Introduction of medium-range ballistic missiles into Cuba (1962)

- Threatening to enter the Arab-Israeli War (1973)

- Supporting the Vietnam armistice violation (1975)

- Introducing combat troops into Cuba (1976-77)

- Supporting Cuban intervention in Black Africa (1975-80)

- Introduction of long-range theater ballistic missiles (SS-20) aimed at West Europe (1977-78)

- Invasion of Afghanistan (1979-80)

It should be quite evident that the Soviets do have the propensity to initiate significant posture changes by surprise. Their social system permits them to act relatively fast and in secrecy; any limitation even distantly resembling the U.S. War Powers act would be unthinkable. As to public opinion, to the extent that it has any basis of fact, it is not in position to influence the Soviet government decision in matters of foreign policy. Be that as it may, the remarkable trend apparent from such a list is that in the 1960 to 1980 period the U.S. response capability in terms of posture changes has *gradually eroded to the point where it hardly seems to influence the thinking of the Soviet decisionmakers.* We are still struggling with the appropriate moves to counter the SS-20 deployment in the European theater; our military options in regard to Afghanistan being (admittedly) unpromising, we have elected to act by economic moves and by the symbolic boycott of the 1980 Olympic games—and even these relatively mild actions have seen their effect diluted and delayed by the clear reluctance of the U.S. Allies to follow suit.

The Soviets have at their disposal a large number of initiatives, all having considerable damage potential to the Western strategic posture, all being susceptible of implementation within relatively short time spans:

• Modify peripheral nations' political alignment via "peaceful means" (Iran, Pakistan, Yugoslavia)

• Develop an antisatellite capability based on radiation weapons

• Reveal or demonstrate step increase in antiballistic missile or antisubmarine warfare operational capabilities

• Break out from under any particular constraints of SALT provisions, related protocols or understandings; e.g., redeploy the Backfire bomber or demonstrate substantially increased range by means of refueling

• Ultimatum, followed by theater nuclear counterforce attack against NATA military assets

• Modify ballistic missile carrying submarine patrol areas and patterns

• Evacuate cities, button down and harden industry

• Launch of central nuclear counterforce attack

These examples have been deliberately chosen to reflect widely different time scales; some of them would last several years and possibly decades; some others may take place in a matter of minutes; they also show actions of widely different magnitude and import in terms of posture modifications. The point here is not to speculate on the relative plausibility, but to realize that they are all within the realm of possibility; they all involve Soviet initiatives, whereagainst the U.S. and more generally the West have no serious deterrent or compensatory posture changes available short of threatening, in final analysis, general and largely uncontrolled central nuclear war.

The cumulative costs to provide the U.S. and other Western nations in a constant state of readiness against all such possible eventualities are prohibitive, especially when seen against a background of endemic state of cold war and confrontation lasting perhaps several decades and when a large number of immediately present, fast risetime contingencies would require massive front-end investments.

The principle which may greatly alleviate this burden is to direct increased efforts toward improving the time rate of change of the Western posture. In other words, increase our capability to evolve relatively fast responses (or initiatives, for that matter), in addition to the currently planned force structure enhancements.

A few examples will help in forming a more concrete grasp of this principle. Combination and redeployment of military assets in the current inventory are probably the most rapid posture changes possible to implement at the present time.[29] It should be possible to develop equipment for dormant storage requiring minimal operation and maintenance costs, even for lengthy periods before use or refurbishing. Facilities for production and transportation may be installed, used for civilian purposes prior to actual need, but rapidly convertible to one or several of their national security related uses. In all these, the "software" associated with the proposed operations must be developed, continuously verified, or even exercised; it includes the establishment of plans, procedures, training of cadres and personnel—all with adequate active or passive protection and security provisions.

Providing for the capability of rapid posture changes (as contrasted to concentrating all our resources on forces in being) is not an inexpensive proposition. By allowing to focus timely operational readiness to enhance our posture in the specific area of provocation or challenge, it may be a more effective way of increasing the worth of our defense-

---

29 The tactical moves on the battlefield or the changes in electromagnetic warfare tactics are (rather arbitrarily) considered under the multiple complexions discussed earlier.

related investment. But well beyond the considerations of investment or cost effectiveness, by paying attention to the dynamics of posture changes, we may lend credibility to arms control provisions which otherwise would not be acceptable to our military experts and to the less trusting segment of the U.S. Senate. By providing for specific insurance-type development and implementation programs which offer timely and effective compensatory response to possible treaty violation, the security of the U.S. would be factually and perceptually enhanced.

One should not be led to believe that increasing the emphasis on the dynamic aspects of our posture development will solve most of the problems; the whole subject has not yet been assessed with the required analytical comprehensiveness and rigor. But, in first approximation, it seems to add to our capability of prehostilities posture management and thereby enhance our deterrent capability. As such, it should be further explored in the context discussed here.

Conversion of Civilian Assets

All Western nations, notably including the United States, face continuing competition between the public funds spent on military readiness ("guns") and those spent for social pursuits ("butter"). The two are competing only in the bookkeeping sense; we have argued earlier that the social benefits and blessings would disappear fast if the external security of the nation is threatened. Conversely, no military power can remain viable and robust (in a democracy, that is) unless vigorously supported by a stable and cohesive sociopolitical system, ultimately based on the consent of the population.

The arguments about the right level of GNP share for military expenditures, about the wastefulness of military procurement practices, and about the abuses of the welfare system or the laxity of government administration in other civilian-oriented pursuits will be with us probably as long as the Republic endures; they seem to be intrinsically part of any democratic form of government.

The conflict between civilian- and military-oriented expenditures may be significantly alleviated, though, if we can find a way to invest in facilities and equipment which in peacetime can be applied to socially and economically justified endeavors, and which in wartime can be rapidly transformed into assets of immediate military value. The capital investments and the peacetime operation and maintenance costs, including those of continuous readiness and training, and successive upgradings, are justified and paid for by the civilian role. These may even show a profit, especially when the operations are under the management of private industry. Other civilian benefits may accrue, such as using the training and readiness aspect of potential conversion programs for job training and skill development related endeavors.

The investment and recurring costs charged against the military budget would include essentially the specialized conversion facilities and equipment, the training and the logistics systems required by the conversion process, as well as the security and protection aspects.

There is no need to argue here the principle involved in the convertibility. Even the Soviets show us an example and some historical precedents go back as far as World War I and probably well before.[30] The U.S. military has endorsed the conversion concept, in particular for the conversion of the Merchant Marine and the civilian reserve air fleet (CRAF); civilian communication and utility systems are part of the U.S. wartime emergency plans.

The only points which should be made in regard to conversion are related to the broadening of the sphere of application and to the need for active wartime protection of

---

[30] The whole Soviet national air transportation system, Aeroflot, is designed and operated with the specific objective, and the demonstrated capability, of supporting military airlift requirements. The French expedient of using Paris buses and taxicabs at the Battle of the Marne (1914); the British flotilla of yachts at Dunkirk (1940) are considered classics.

the assets involved. If, for instance, merchant vessels are to play a role in sealift operations, they must be provided with autonomous air defense and antisubmarine capabilities consistent with their value as targets for the enemy, so as to alleviate the task of convoy escort incumbent on the Navy. A similar argument could be made for the airlift component and the associated civilian air terminal facilities.

Commingling

The title chosen for this section is intended to convey the general concept of exploiting the possible beneficial inter-actions among mission needs, system performance charac-teristics, technology, program management, deployment, training, operation, and maintenance. Taking the U.S. Department of Defense as an example, the administration and the management of resource development is organized according to Service Department lines. Mission needs, evolved jointly by the operationally oriented user commands and the cognizant Service staffs are eventually assigned to individual Services for development and procurement, fol-lowed by test and evaluation prior to operational deployment. Technology and exploratory development programs are also under the management of the individual Service Departments. The compartmentation implied, and in fact caused, by this management organization may have helped in the early 1960's in the assignment of responsibility and accountability, but is no longer justified in the present and foreseeable environment, nor can it be viewed as an outstanding contributor to the much-touted cost effective-ness of the defense establishment.

Already now, but most certainly in longer-term future, all major missions are seen as *interdependent* for the achieve-ment of their respective objectives; they are furthermore *interconnected* by many essential facets of combat scenarios. For instance, overseas force projection, including the Rapid Deployment Force concepts, is not an attractive proposition, unless local air superiority can be assured against the Soviet-supported enemy. As another instance,

Fleet operations are unthinkable in presence of long-range air-based threat without the cooperation of long-range air surveillance; protection against the long-range submarine-launched cruise missile threat is greatly enhanced by fixed ASW detection and localization systems. As an example of essential combat-scenario-based interconnection, the role of central nuclear war capability and readiness should be cited as controlling to a significant extent the evolution of theater-level conflicts; conversely, the course of theater war may be the decisive element in the escalation to the central nuclear war level.

Many systems can potentially contribute to several mission objectives; thus a long-range bomber with the appropriate payload complement may serve as a competent maritime patrol and antiship aircraft. Airborne early warning may also serve as an acquisition sensor for ballistic missile defense; evolution and netting of ground-based air defense may eventually lead to respectable antiballistic missile defense capability.

That the entitlement to certain missions of any one of the Service Departments is no longer essentially justified (except as supported by the engrained "roles and missions" traditions) is evidenced by the organization of numerous Joint Commands for operational mission responsibilities, as well as for training and readiness. Increasingly, these Joint or Tri-Service-type organizations also assume responsibilities for requirement definition, and in many instances, for system development.

Many systems, associated with often widely different missions, share technology applications, components, subsystems, and ancillary and/or support systems. Navigation satellites are an outstanding example of this trend.

We should also notice that the sharp delineation between nuclear and non-nuclear warfare will progressively disappear, at least in military doctrine and procedures, if not in the public mind. Already it is difficult to distinguish the boundary between highly effective "conventional" battlefield weapons and small-yield nuclear weapons; in the

ABM defense context, conventional warhead-carrying missiles are considered as a serious threat to nuclear missiles aimed at ICBM launchers; conversely, it is possible to think in terms of non-nuclear lethal effects to incapacitate major nuclear weapon bases, both on land and at sea. The commingling of nuclear weapons and those with conventional warheads with no "functionally related observable differences" [31] are a matter of consideration in arms limitation talks as well as in security problems associated with theater nuclear wars. The distance of the target area from the weapon bases is no longer a valid criterion. Central nuclear war oriented (and SALT controlled) launchers may in fact be used in theater context (they are so used now) and "gray area weapons" are definitely a factor in the central nuclear balance.

In view of all the above, and unless major valid objections are raised against the "commingling" principle as defined here, procedural means should be explored for the purpose of exploiting its advantages in future defense related procurement and operations. The immediate advantages, beyond the simple avoidance of duplication, are connected with the previously discussed "multiple complexions" and "dynamics of posture changes." The overall costs and concept-to-SOC[32] time spans may also be reduced in most gratifying proportions.

Information War at the Force Structure Level

The information war is a broad concept covering basically all the attempts at degrading or destroying the information available to the enemy to conduct his operations, and conversely to protect the information flow supporting the friendly forces from similar undertakings.[49]

---

[31] "FROD's" in the arms controller's parlance.

[32] Significant Operational Capability, usually interpreted as 25% of the planned force levels in the operational inventory.

In various forms the information war concept has been with us for many centuries.   In point of fact, it has been referenced by no less person than Sir Winston Churchill under the expression "Wizard War."[23] The Soviets also have under various names published extensively on the subject, especially within the last decade, and lay considerable stress on the proficiency of their military forces in its operational applications.   It has also received very much increased attention in the U.S. as of late, mostly in the overt military context, owing to the substantial progress in electronic data processing technology[33] and to its much increased importance in interacting with complex and smart weaponry, as well as with the extensive military communication networks.

The best known immediate applications of information war relate to *tactical* engagements, but the concept should not be limited to that level.   Specifically, it should not be restricted to electronic countermeasures (ECM), counter-countermeasures (ECCM), or to attacks on command and control systems, ("Counter-C^3").   In fact, we have just seen applications to weapon system basing; the multiple complexions, say in weapon deployment, by saturating the enemy's targeting decision process implement a typical move in a more general aspect of the information war.   Still at the level of military operations, elements of the command and control system can be used in a nondestructive and reversible mode by serving as warning or deceptive message generators.   These too can be seen as elements of an ongoing information war, intimately intertwined with other, better advertised, aspects of warfare; they could be designated as the *"strategic"*-level elements.

At the level beyond and above the conduct of military operations and in a quite different time domain, principles related to the information war concept must be considered in relation to the development, deployment, and performance characteristics of weapon systems and their ancillaries.   These principles are conveniently referred to as *"controlled visibility"* of our military forces; this designation

[33] Appendix—Note U.

has the advantage of conveying more than simple "security." This level of the information war is seen as related to the *force structure* level. What is really meant is not only the attempts at information denial, but the deliberate generation, formatting, and transmittal of message streams containing a time-variable mixture of true and (logically consistent but) false information, augmented by appropriate doses of neutral "noise"—the whole intended to increase the workload of enemy intelligence analysts *and* to decrease their own and their superiors' confidence in the reliability of their conclusions.

Recapitulating now, three levels of the information war concept have been identified:

- The *tactical* level, aimed at the functional characteristics of specific engagements, such as a missile intercept of an aircraft or an air-superiority battle between two opposing squadrons;

- The *strategic* level, aimed at degrading the enemy strategic decisions, such as order of battle, preemptive moves, commitment of resources. Not incidentally, the appropriate "messages" conveyed in this context may carry the essential elements of deterrence, as we have seen, through fear, anxiety, or doubt;

- The *force structure* level, aimed at degrading the enemy's views on our current and future military capability and readiness. Again, the dual objective here is to elicit the wrong development and procurement decisions by the adversaries and, through the appropriate (in this case, long rise-time) signals, strengthen the deterrent aspect of our force structure.

This section is more particularly aimed at the force structure level.

A few specific illustrations (described, for obvious reasons, in general terms only) will help to grasp in more concrete terms the concepts involved. Test programs, exercises, field maneuvers, and routine deployment/redeployment operations should be planned and conducted[34] with appropriate consideration of the message content as constituting a possible move in the information war. One may even start , thinking in terms of designing peacetime military activity with the dual purpose of increasing our proficiency while offering controlled visibility to the enemy.

The ostensibly described "mission element need statement" (MENS) may satisfy the orderly program planning and budgeting requirements and, to that extent, must be in reasonable conformance with the truth. Whether or not it needs to be the whole truth, at least at the relatively low security level of the MENS, is left to the reader's appreciation. It may be possible to develop and exercise force components for specific and well-circumscribed mission needs, then—concurrently or subsequently, deliberately or fortuitously—discover other mission elements also requiring the corresponding capability. The "commingling" and "conversion" principles offer a few intriguing opportunities in this direction.

Along the same line of thought, it may be possible to plan and conduct development programs (or at least generate credible and realistic observables) aimed at misleading the enemy's own research, development, and procurement decisions. There is no clear evidence that the Soviets really use (at the present time, anyway) the "U.S. threat" as the basis for their force structure planning, but if our currently advocated endeavors truly bear fruits, i.e., if the U.S. recovers its initiative and technical leadership, they just may start to listen and ponder again.[35]

---

34 (Note added in proof.) No one would deny in these days that the Warsaw Pact spring maneuvers in March-April 1981 exert strong influence on the evolution of events in Poland and on the reactions in the West.

35 Whatever the outcome of this type of activity might be, the U.S. development programs must assume that the enemy eventually will have perfect information on the characteristics of our own weapons and ancillaries; that they will develop plans for countering whatever our corresponding incremental capability might be; and that they will have the technology and production base to field their "responsive threat" in time to defeat our move. Considering the relatively ponderous procedures in the West, this

As repeatedly pointed out, whenever any information war move is being contemplated, we must at the same time examine our own vulnerability to similar moves by the enemy. As the strategic and force structure level moves exhibit increased degree of complexity and are protected by increasingly stringent security compartmentation, attention must also be paid to the risk of deceiving or manipulating our own selves.

When the problem of controlling the visibility or, more generally, the information stream related to our force structure, is being discussed, the constraints imposed by the nature of a "free and open society" invariably come to the fore. Should the freedom of information principle cover the mission objectives, general capability and programmatic aspects of our force structure? How do we ensure that our defense dollars are not wasted? Can we trust the military judgment in defining future requirements? Questions of this type lead to continuous, high intensity level public discussion of our military capability, strategies, tactics, results, and day-to-day events. Senior and junior military officers are routinely summoned to testify in public hearings about cost overruns, performance deficiencies, and schedule slides; the cosmetic deletions in the Congressional Record, made in the name of security, may elicit a chuckle among knowledgeable experts of the Soviet intelligence establishment, but certainly do not impede their efforts aimed at correct interpretation. The dogged persistence of the recently discovered investigative reporting almost puts a premium on tracking down and advertising weaknesses and mistakes. It is far from certain that the related court rulings have given full weight to national security considerations. The strict attention given to budgetary

35 (Footnote continued)

idea is not really far-fetched. In our own development and procurement plans we should specifically include considerations of fully-informed responsive threat growth. As the system characteristics are bound to eventually leak out, and as the enemy responds by modifying his own offensive or defense threats, we should have mapped out a whole series of appropriate future growth responses prior to the commitment of the weapon system to development. I specifically suggest that the consideration of "time window for operational effectiveness" should become a factor in procurement decisions.

controls forces the military establishment to state forcefully and perhaps somewhat overstate, but certainly to
publicize, the weaknesses in our current and projected
posture.   This is particularly irksome in the context of
strategic nuclear forces; the situation there is almost
pathetic.[36]

The information flow related to the specifics of our military
capability should be controlled.   Congress has all the
authority to receive accurate statements under rather
strong constraints of secrecy before it reaches the adversaries, the neutrals, the Allies, and the U.S. public.   The
freedom of information act should definitely not apply to
the United States and the Allied military potential.
Whenever there is a reasonable probability that sensitive
information might leak out, it must be surreptitiously
compensated by a plausible volume of deliberately generated confusing and saturating messages.   All the details
related to this aspect of the information war should be
protected by security classification at the same level as our
intelligence operations. The concept of manipulating enemy
public opinion and military decisions in the context of arms
control is discussed in more detail elsewhere in this study.[37]

The principle comprising the deliberate manipulation and
control of enemy visibility on our military force structure

---

[36]   There is a built-in positive feedback in this process. As we observe the
growing rate of Soviet strategic investments, our military seek more support from Congress. This is a place of public debate and senior officials,
supported by the defense industry and think-tank analysts, must resort to
vigorous criticism of our currently planned posture. They basically contend with the appropriate vigor, and often in tones of despair, that we are
weak and we are likely to be defeated if the present trends are allowed to
continue. The Soviets, of course, capture the gist of these debates and
enthusiastically present to the Politboro our own conclusions as evidence
that their investment has paid off. The result is that they are encouraged
to continue at a high level of investment while our defense establishment
must be, more often than not, satisfied with piddling funding increments,
grudgingly bestowed, for still another study, or interminable analysis of
alternative options. The process continues until such time as our weakness,
from being just conjectural projection, becomes a real, tangible and
irreversible fact.

[37]   Appendix—Note F, and Appendix—Note V.

and capability should be given increased attention in the future. It may assume new dimensions as other Western nations come to share the responsibilities, as well as the benefits, of a much improved military posture.

* * *

Irrespective of any set of powerful general principles being applied, the recovery of effectively usable military power for the West requires resources beyond the reasonable limits of long-term U.S. commitments. Responsibilities, burdens, and benefits should be equitably shared among the Western nations and their major trading partners.

The principles just discussed, plus perhaps many others yet to be discovered, may help considerably in improving the effectiveness and the end product of our military posture development process. We should, however, not delude ourselves into believing that the corresponding financial burden will be much lightened. No less than the development or the reacquisition of a world-wide military capability is needed, which should within the 10 to 15 years reliably satisfy the projected security requirements of the Western nations, including possible new accretions to the Western Commonwealth.[38] This military capability, even in its initial evolutionary phases, may have to contend with all the machinations and explicit military threats of the Soviet Union and of its cohorts, should they feel threatened anew by our evolving strength and should they persist in following a path of continuing struggle and confrontation.

The post World War II military investment picture has been grossly distorted by the U.S. self-image as the leading power of the West and by the economic realities of the 1945-1960 period. Our military budgets, stimulated by the Korean conflict, have dwarfed in both relative and absolute

[38] Strong and reliable military power may be the most effective attraction for potential candidates, at least in the initial phases when ideology must be of necessity tempered with common-sense concern for immediate security.

terms the aggregate defense-related investments of all Western nations. [39] Our institutional military R&D momentum has correspondingly grown in magnitude and sophistication far beyond those of our Allies and, for awhile, beyond those of the Soviet Union as well. For reasons variously ascribed to generous altruism, to enlightened self-interest, or to gross attempts at imperialistic exploitation, we also supported the Allied (and former enemies') recovery with economic assistance under the Marshall Plan and its sequels. Cumulative expenditures over the two past World War II decades under the heading of mutual assistance appear staggering in comparison with the current U.S. defense budget items at issue.

When envisioning the future plans for the development of true, competent Western military power, one must start with the firm apperception that this picture is by now definitely and irreversibly altered. The GNP's of Western Europe exceed, that of Japan approaches, the U.S. figure; the Western nations' populations, in the aggregate, out-number the U.S. approximately 3.5 to 1; the major population centers, which are possible candidates for joining the West (Brazil, Mexico, and India) would raise this projection to 5:1 by 1990, with the average GNP/capita quite compara-ble to that enjoyed by the U.S. There is no economic reason why the U.S. military investments should be any longer grossly out of proportion with those of potential Common-wealth partners. In terms of technology, a few nuclear weapon related areas excepted, the European, South African, and Israeli capability is quite equivalent to that of the U.S.; Japan has the potential, but not (yet) the resolve, to follow suit.

The problem of providing effective military power in support of the community of Western nations will not be solved through greater exertions by any one, or by a small number, of member-nations acting in isolation or in ephemeral alliances, contracted or dissolved on an ad-hoc basis as their respectively perceived short-term interest

---

[39] Israel and South Korea are the exceptions, but again, the actual technical and financial resources were mostly provided by the U.S. under a combina-tion of support mechanisms, euphemistically termed "mutual assistance."

may dictate. It must be addressed on a permanent, long-term basis, supported by mutually agreed principles of equity and effectiveness.

The concept of the Western Commonwealth, as described in Chapter XI, may offer a framework to formulate such principles. It should be based on a full measure of mutual cooperation in matters of common defense, aimed at ensuring the security of the community as a whole as well as the security of individual nations. Some of the more immediately apparent implementation aspects are being explored, but it is quite evident that the essential task of establishing the viability of a truly integrated Western military defense structure still remains to be accomplished. No discussion of military power would be taken seriously if it fails to mention the paltry considerations of financial resources—money, in short.[40] The Western nations have ample resources and one may argue at length about the exact GNP fraction (always a small one) that would be appropriate for defense. The brutal, unmistakable fact is, however, that the integrity of military power should not be jeopardized by the confluent pressures of domestic social goals and by the resource drain primarily imposed by the short-term interests of the oil producers. One conclusion is quite clear from all the preceding chapters: While the Western democracies have a vital interest in providing for effective military power, the interest of their resource-rich trading partners, (specifically the newly emerged petropowers), is hardly less vital; in point of fact, it is more immediate. Instead of (or in addition to) the West "selling" or giving military equipment to the insignificant national defense forces of the petropowers, we should endeavor to conclude long-term financial agreements, whereby they would in effect underwrite a significant portion of the Western military budgets in exchange for firm guarantees of military support and

---

[40] It is instructive to reflect on the thought that the British Parliament would have never risen to its political preeminence, except for the dire needs of the Tudor and the Stuart rulers to "support the glory of the King" with ample tax levies, voluntary donations, etc.; precursors to Congressional defense appropriations and authorizations. Napoleon I (still in the ascending phase of his military career) allowed that "l'argent, c'est le nerf de la guerre"...

cooperation. The basis of this suggestion is that, irrespective of religious or ideological differences, the future well-being of the petropowers, especially those at the immediate periphery of Soviet influence in the Middle East and the Caribbean, is indissolubly tied to the security and prosperity of the West.

*

This is a rather complex and lengthy chapter; perhaps it is time for us to pause and to ponder.

The probability of war is here to stay, well beyond the limits of our time horizon. Without effectively usable military power, all the other slowly evolving processes implied in our first three premises will not be given a chance to succeed. The unrelenting Soviet military thrust is the primary driving force behind the urgency of making effectively usable military power available again to the West. The massive and steadily growing power of the Soviet forces-in-being, coupled with technology advances in the fields of targeting, weapon range, accuracy, and kill mechanisms seriously threaten the viability of many essential military missions, heretofore taken for granted as being available in support of Western foreign policy objectives.

Instead of continuing to simply argue for increased defense budgets, a certain number of principles might, if properly applied, enhance the effectiveness of defense-related investments.

Multiple complexions in the basing of military assets are seen as a way to enhance deterrence and fighting ability of military forces which can not, by policy limitation, take the initiative for military action. This principle has been applied to strategic nuclear forces; it should be expanded to theater nuclear weapons, overseas force projection, space-related assets, naval warfare and, most emphatically, all command and control, as well as logistics systems.

Increased attention to the dynamics of posture changes may lead to more responsive and more specifically focused reactions to hostile initiatives. Deterrence and fighting ability are thereby improved and the dangers of over-reaction decreased. Investment in the "time derivatives" of the force posture (including military, civilian, and industrial aspects) is suggested. The conversion of civilian assets for military use appears important from the viewpoint of effective utilization of resources at the national level. It may also help the dynamics of posture changes and socially desirable by-products in terms of facilities construction, skill development, and stockpiling of equipment and raw materials.

Many missions are interacting with each other and are, in fact, interdependent. Technology has often multiple applications and many systems, once developed, may contribute individually or in combination to several mission objectives. Joint benefits derived from the "commingling" of mission elements, as well as system component developments, appear to be potentially promising.

Applications of ideas derived from the higher reaches of the information war concept may be helpful in alleviating the problems of an open society which otherwise fully and effectively signals to the enemy not only its current force structure, but also its capabilities and future development plans.

The financing of a step increase in the Western defense related investments requires a definite broadening of the available resource base. Apportionment among all the Western nations, the potential beneficiaries, is a step in the right direction; levying contributions, in the form of long-term financial arrangements on our resource-rich trading partners, may point the way to a more satisfactory solution.

*

If it be true that good research results in more new questions than new answers to old ones, then the work reported here must qualify as clearly outstanding. A number of interesting and potentially important questions have arisen in the course of our explorations, but little more can be done here than bring them up as challenges to the reader and to future contributors.

Is it really desirable that the West shift the bulk of its surface and airbased offense weaponry in the direction of "smart" missiles?   The experienced military distrust, probably on excellent empirical grounds, sophsticated high-technology hardware, which performs in an outstanding manner under controlled conditions but is apt to fail catastrophically in the havoc of the battlefield. If there is one result to be retained from their long and arduous training, it is that the only software which they can trust for easy and adaptive reprogramming under unpredictable battle conditions is neatly packaged between their own ears and between those of their own trusted subordinates. Is it possible that other approaches, perhaps based on a variable mixture of "smart" weapons and of massive, low-cost fire-power, less sensitive to the specific battle conditions, might be a more attractive solution?   If the "neutron bomb" (enhanced radiation tactical warhead) becomes indeed part of the Western inventory, would the emphasis on precision weapons still be justified?

If the armed forces of the Western democracies are to support Allies from the Third World, is our force posture truly oriented toward enhancing their autonomous combat capability, or are we constrained to explicitly taking part in the fighting, simply because our weapons and ancillaries are not effectively transferable?

Is the proliferation of nuclear weapons (specifically, battle-field-level nuclear warheads and relatively low-yield strike weapons, with limited delivery ranges) truly the unmitigated evil implied by our current policy?   Whatever the moral issue, proliferation is a fact of life which will not go away; it constrains only the friends and clients of the West, or those which the Soviet Union, for reasons of its own, wishes to keep constrained.   All others will in the next couple of decades have rudimentary (by U.S. standards) but quite workable nuclear armaments. What are the true interests of the West in this matter?

Lastly, based on the observations of this study, it is quite apparent that a number of tactical disadvantages result from our policy restrictions in regard to military initiatives. There seems to be an evil connotation of being the "aggressor." A real crisis, leading to military action, always implies several, often confused, steps where the moral responsibility for aggression is hopelessly buried in the micro-structure of the scenario. Perhaps thinking along the lines of "soft kill" (actions aimed at degrading the enemy's sensor and command and control systems) might go a long way toward further reducing the risks associated with the current, generally applicable rules of engagement which clearly discourage *our* offensive first moves.

*

One may, or one may not, accept our fourth premise related to military power; many of those engaged in thinking about the problems discussed here may disagree with the implications or conclusions of this chapter.   There will be unanimous agreement, however, to the effect that a lot still remains to be done; some may even feel that this is quite an understatement.

CHAPTER XIV
Interactions

THE POLICY OBJECTIVES DERIVED FROM THE MODI-
FIED PREMISES CAN NOT BE IMPLEMENTED IN ISOLA-
TION. JUST AS THE COMPONENTS OF THE SHIFT IN THE
ENVIRONMENT, THEY MUTUALLY IMPACT AND CON-
STRAIN EACH OTHER, BUT THEY ALSO OFFER THE
POTENTIAL FOR POWERFUL SYNERGIES. THE DEVEL-
OPMENT OF A COHERENT SET OF POLICIES MUST BE
LEFT TO THE POLITICAL PROCESS SUPPORTED, RATHER
THAN REPLACED OR PREEMPTED, BY ANALYSIS.

For the sole purpose of orderly discussion, an attempt was
made to dissect the shifts in the societal environment into
separate components (PART 2): information flow, social
stresses, ecology, economy, and military. Artful truncations
had to be contrived in order to maintain even a semblance
of separateness, and all this to no avail: The various aspects
of environmental changes are hopelessly intermingled. The
hope for an orderly discussion of mutually independent
policy objectives (Chapters X through XIII) is equally fated
to disappointment; if anything, they are more intensively
interacting with each other than the environmental
components. This mercifully short chapter discusses some
of the characteristics of the most conspicuous interactions.

The preceding chapters undoubtedly justify the notion that
we are dealing here with problems of the most unusual
complexity. (I) The primary interactions between the four
premises [1] are by definition mutually supportive, but they
can be strengthened or weakened by secondary interactions

1   Chapter IX.

between the implementation aspects of the derived policy objectives; (2) Further changes in the environment may modify the nature or the impact of interactions; (3) Many interactions, especially at the lower implementation level are unknown at the time of policy formulation; their impacts can not be foreseen with any high degree of confidence and may be found irreversible by the time their effects are observed. A few examples will help to illustrate these points. Since most of them have been discussed in earlier chapters, only the interactive aspects are emphasized here.

- ## Control of Information Flow vs. Free Speech

  The intense and pervasive information environment contributes to the undesirable social and economic turbulence caused by single-interest pressure groups. Purported remedies are being forced upon the public before their longer term impacts are being ascertained; this creates unnecessary stresses and dissipates much of our resources and enthusiasm. In another dimension, the easy access to communication channels allows hostile manipulation of public reactions by those who have interest in doing so. The prisoner-of-war issue in Vietnam (1971-73) and the Iranian hostage crisis (January 1981) are classical examples.

  Here the observed detrimental effects are to be weighted against the dangers of inhibiting free speech and other Constitutional guaranties. There is also a valid argument to be made in favor of "free and full debate," characteristic of democracies, which ultimately leads to the adjudication of issues to the best interest of the public. Unfortunately, basic data is lacking on the effectiveness and the dynamics of this process; it is also fair to confess that the relationships between societal communications and collective political behavior are at best empirical and putative at this point of our knowledge. Attempting to formulate policies in the absence of basic data is fraught with risks; some would argue that in such cases it may be wiser not to have a policy. Decisions are made,

however, every day; in the absence of any other reference framework, the political process takes its course
and follows the pressures of immediate expediency as
determined by the voters. One could, in general, do
worse.

- ## Domestic Transfer Payments vs. International Competitiveness

Since World War II, the increased emphasis on "social
justice" is being reflected, among other moves, in a
substantial growth of "transfer payments" with the
openly stated purpose of income redistribution in favor
of the underprivileged. Differences in social legislation
and benefits, coupled with other aspects of worker
productivity result in highly perceptible differences in
competitiveness among the Western nations. The issue
here is the age-old classical conflict between the near-
term concerns of the poor and the longer time horizons
of the affluent. The former are not always genuine and
justified; the latter are hardly ever free from selfish
greed; the balance between the two is a matter of
personal philosophy and vantage point.

In this case, the tying of social benefits to productivity
(a policy derived from the first premise) is seen to
support a policy objective associated with the second
premise, namely competition on advantageous terms
among the industrialized societies.

- ## Protection of the Environment vs. Growth and Stability

The intensifying concern for protecting the physical
environment diverts capital investment from being
primarily aimed at near-term profitability, reduces the
nation's economic growth rate, eliminates certain jobs
or industries, increases the "social overhead" and thus
contributes to inflation. There is no possible argument
when the issue is as clear cut as the elimination of
toxic wastes in drinking water or the protection of the
ecology and scenic beauty of the seashore against oil
pollution. But when (say) the nuclear power industry is

being attacked not on the basis of safety and cost, but by raising the specter of the largely unsubstantiated radiation hazard associated with normal operations, potentially affecting generations as of yet unborn, value judgments are difficult, dogmatic statements abound, and the answers are invariably controversial.

Again in this example the long-term environmental concerns compete with the immediate need to generate the substance which will in fact offer the basis for any "quality-of-life" to a growing population with growing expectations.

- ## Military Preparedness and Reindustrialization

The resources directed toward reestablishing an adequate military posture for the West can be channeled to support modernization and relocation of a sizeable fraction of the manufacturing industries. Pursuing the concept of societal resilience (the ability to withstand major perturbations, including nuclear combat environment), a number of infrastructural investments and stockpiling of finished and semi-finished products can lend a measure of stability to otherwise highly cyclical industries.

Given that resilience is a desirable characteristic for any society in a world exposed to natural or manmade disasters; given that industrial production and employment are, in a system dedicated to free enterprise, subject to most undesirable fluctuations; finding approaches which alleviate problems in one area, while clearly helping the other at the same time is "obviously" attractive. Even in such cases vigorous challenges to the synergistic approach might (and probably will) be voiced. The arguments may well be in valid regard to competition with private enterprise for capital, raw materials, and labor; known or suspected inefficiencies in any government-supervised large-scale undertaking and, perhaps, on the basic issue whether or not business cycles are beneficial in the long term by eliminating the least efficient competitors.

● Military Cooperation vs. Technology Transfer

Military-related high technology and its extension, space technology, have beneficial fallouts toward the creation of new industrial sectors and increased competitive leverage in international trade. The problem of deliberate or inadvertent transfer of key technologies, concomitant to military cooperation, creates continuing tensions within the Western alliance.

This is an illustration of the fact that a given desirable policy objective (military cooperation) can lead to undesirable side effects. The nation or the industry which has the technology lead in a specific situation is deprived of the reward it would normally reap in a purely civilian area of endeavor. The recipient foreign nation or industrial sector may use the newly acquired technology to compete aggressively against Western interests.

● Interdependence with the LDC's

The high level of per capita resource consumption, being (perhaps unwisely) accepted as an element of affluence and quality of life in the U.S., forces the nation toward interdependence with the LDC's. This in turn justifies in the eyes of many the "entangling alliances" our first President so eloquently warned against. But then, this interdependence and entanglement may also lead to potentially promising nonviolent means for protecting the emerging nations against the ideological onslaught of communism.

This type of interaction is illustrative of difficulties encountered in the process of policy formulation: none of the benefits are indisputable, nor are the ascribed evil consequences unmitigated. Other controversies arise along the near-term vs. long-term dichotomy, still others about the true identify of the beneficiary classes.

- ### Agricultural Research: Domestic vs. LDC Applications

  Research and Development programs originally undertaken and justified on the grounds of helping domestic
  agriculture and industry, may pay off in improving the
  balance of trade, but it may also offer unexcelled
  opportunities to contribute to the progress of less-
  developed nations. U.S. research in plant hybridization
  stands as a shining example of foresight in the 1920's;
  one would love to describe it as a deliberately undertaken long-term move, but the truth is that it was
  mostly driven by the enthusiastic competency of a
  small number of researchers.

  The conflict here may be seen as typical between the
  near-term and long-term considerations. By increasing
  our own food surplus, our balance of payments and
  political leverage are improved. By introducing intensive agriculture into the less developed regions, we
  increase the need for fuel, fertilizer, and machinery.
  This in turn depletes the resources badly needed to
  support the growing population (habitation, schools,
  infrastructure) which unavoidably follows the increased
  food supply. The current conclusion seems to suggest
  that agricultural developments should go hand in hand
  with education and family planning—a conclusion more
  often than not vigorously resisted by the potential
  beneficiaries.

- ### Abrupt Changes in Trade Gradients

  In matters of international trade, the intensified
  communications flow causes commodity and cost
  gradients to disappear unless based on some intrinsic
  resource availability. This leads to shifts in the balance
  of trade, in some cases even to successive reversals,
  often encouraged and accelerated by the purposeful and
  aggressive investment policies of at least one of the
  participants.[2]

---

[2] Within the last 25 years, the U.S. moved from a position of world-wide
domination of the automobile industry to the point that we have to consider trade barriers to protect our domestic production against the Japanese and European imports.

The rapid increase in Mexico's proven oil reserves already has changed the current and potential position of that country vs. the U.S. and other industrialized nations. If the deep-strata natural gas extraction justifies its promise (and some experts think that this is a very real possibility), it is to be expected that the role of Mexico, with its growing population, may well become decisive in the balance of power within the Western hemisphere.

Unexpected discoveries may shift international trading patterns within a matter of a few years. Norway, for instance, known for centuries as having its national economy based on shipbuilding, shipping, and fishing (with pulpwood and tourism added in more recent years), has suddenly entered an era of new prosperity and influence in the wake of the North Sea oil discoveries. The traditional pursuits of seafaring and fishing are said to have experienced sharp declines in terms of share in the national product.

Gradients may also vanish or appear for purely technological reasons. For instance, if nuclear fusion power becomes industrial reality (as some expect) within a matter of a decade or two the future of fossil fuels, and by repercussion that of the oil-producing nations, will be seriously impacted.

• Political Alignments vs. Environmental Concerns[3]

In March-April 1981 the public concern about nuclear power plants in West Germany (FRG) has erupted in a series of violent protests. Students, as usual, were in the forefront; their feelings and arguments are essentially those which have been debated in the U.S. over the past decade: Nuclear power plants are unsafe, they endanger the environment through radioactive leakage and waste, and are probably not required in view of the alternate potentially available energy sources.

_____
[3] Added in proof.

The energy future of the FRG strongly depends on the oil imports, mostly from the volatile Middle East, and on the natural gas supplies potentially available from Soviet Russia.   If nuclear power developments are curtailed, and if the Middle East remains volatile as it is now, the FRG has no other choice than increasing its dependency on Soviet Russia.   The possible consequences of this trend with respect to the cohesion of NATO are well known by all; they are eagerly anticipated by the Soviet policy makers, just as they pose unresolvable dilemmas for the NATO participants.   The shift in the West German public opinion, on a controversy probably without merit or substance (the desirability of nuclear power plants) may adversely impact one of the major U.S. alliances at a time when such alliances are truly critical.   This in turn may force the U.S. to entertain strategic moves in support of stability in the Middle East—moves which may well dominate our military requirements in the near-term future.

- ### Climate Changes and Aggressiveness in Soviet Foreign Policy

It is reasonably well established that the Earth climate is subject to relatively minute fluctuations caused, at least in part, by changes in atmospheric transparency to solar and infrared radiations.   These changes have well-authenticated man-made components ("anthropogenic"), and some others due to natural causes, such as volcanic activity.[53]   The seasonal average temperature changes are measured in a small number or even a fraction of a degree, but the effect on the growing seasons in the Northern Hemisphere can be surprisingly large.   In marginal cropland areas, such as those found in many parts of Soviet Russia, any significant adverse change in the growth season may be equivalent to one-quarter or even to one-half of the expected harvest being lost; if sustained over a few years, the result could be chronic famine in the affected regions.

There is a plausible case to be made to the effect that if (1) Large-scale crop failures become the norm within

Soviet Russia and cause chronic food shortages; (2) At the same time the world-wide grain harvest, mostly under the control of the Western democracies, remains relatively unaffected, and attempts are being made to exploit this advantage for political purposes; (3) The military supremacy of the Soviet Union is recognized in regard to conventional warfare, while the U.S. deterrence capability of general nuclear war is less evident or more questionable; — — under such circumstances the Soviets could well be tempted, or even forced, to provoke war in order to acquire reliable food supply sources, while achieving simulataneously other political objectives.

This scenario is possible, although its liklihood is not broadly accepted. It serves, however, to illustrate the point that fortuitous natural events (relatively large-scale injection of fine volcanic particulates into the stratosphere) independent of, or superimposed upon, the effects of man-made secular climatic degradation (cumulative increase in the atmospheric carbon dioxide due to industrial activity) can possibly trigger major military confrontations.

● Nuclear Power Plants and Weapon Proliferation

Even though the cost elements are being persistently questioned, nuclear power is an attractive alternative to being subjected to the political and economic indignities associated with the oil imports, if only the safety and pollution problems can be solved to the satisfaction of the public. It also happens that the reprocessing of spent nuclear reactor fuel could yield weapon-grade fissionable material (enriched uranium isotope 238) and an appreciable quantity of plutonium.   Breeder reactors, in fact, deliver plutonium as one of the results of the process.   Both the enriched uranium and the plutonium are ingredients of nuclear warheads.

In spite of treaties and safeguards, any nation that has under its sovereign authority nuclear power plants, or that can cause its spent fuel to be reprocessed under its own authority, has the potential to develop and to fabricate nuclear warheads.

Possession of potential capability to wage nuclear war, no matter at what reduced scale, is being frowned upon by the current members of the "nuclear club." The U.S. seems to be more affected by a sense of responsibility in this regard, due perhaps to the guilt feeling associated with the original sin (or mistake) of having, metaphorically speaking, "let the nuclear genie out of the bottle." Justified or not, there is a basic dilemma between attempting to prevent nuclear weapon proliferation on one hand, while trying on the other hand to encourage the friendly and fuel-hungry nations to become self-sufficient in terms of energy supplies.[54]

- <u>Militarization of Industrialized Societies (West and East)</u>

Since it was not discussed explicitly in the preceding chapters, I have left to the last the long-term consequences of militarization (or perhaps, re-militarization). Since the dawn of history, the emphasis on military power and the specific technologies supporting that power were to a large extent determinant factors in social organization.[4] Without discussing this topic in depth, the possible interaction between advanced technology weapons, technically educated citizens playing the role of military reserve forces, and the need for massive civil defense should be explored as one of the sequels of the present study. While it is possible that

---

[4] The introduction of the phalanx among the warring city-states of ancient Greece as a decisive factor on the battlefield is said to be the basic cause for the evolution of democratic forms of government.[5] During the Middle Ages in Western Europe, the armored knight achieved mastery of the battlefield and, concomitantly, the high political and economic status reflected in the principle of feudalism. Walled cities, being impregnable to knights on horseback, shifted the balance of social affluence and status toward the guilds and the town magistrates, burghers, merchant-adventurers, bankers, and explorers. Not until the late 18th century did artillery begin to dominate the battlefield at the same time as the industrial revolution allowed the mass arming of citizen-soldiers with quite effective firearms, completing the circle of democratic (or socialist) political systems.

such evolution will have some important unforeseen consequences for the West, it is almost certain that it will eventually bring far more drastic changes in the Soviet Union and its satellites. It is simply not possible to subjugate for a significant time period a technically sophisticated population, trained to handle weapons— especially when exposed to communications not subject to effective government censorship.

*

With very little additional effort, the list of these illustrative examples could be much expanded. In point of fact, it is hardly possible to turn to any page of the preceding chapters without detecting cross-impacts between the several suggested policy derivatives. To make it more explicit, it should be clear by now that societies (and, specifically for our discussion, Western societies) do not respond on a one-to-one basis by narrowly circumscribed discrete policy changes to individual stimuli due to environmental shifts. The *ensemble* of all policy changes (including spontaneous initiatives, perturbations, channeling of resources) responds to the multi-dimensional modifications in the environment. Successful adjustment depends, as seen earlier, on the early and adequate perception of the important dimensions of the changes in the environment, followed by the ability to encompass, as part of the policy recommendations, at least the important first and second order cross-impacts.

One obvious conclusion is that policy objectives should be scrutinized not only with respect to their immediate and direct consequences but also in regard to the possible near and longer term interactions ("cross-impacts"). This admittedly is not an easy task and it can not be approached with any reasonable hope of success by hastily assembled teams of specialists, "task forces," or committees. When facing the problem of intelligent choices among policy alternatives and (in the view of many) the far more pressing problem of immediate resource allocations, both methodology and empirical background are simply not available at the national and global scales required for our purpose.

Any approach claiming the mantle of rationality should start by assessing the nature and magnitude of the problem. When attempting to evolve by analytical means a set of policy objectives attractive in the context of societal adaptation, the nature of the problems encountered is most unusual (to the analyst) and the magnitude of the task, while surely underestimated, appears formidable.   As is often the case in "soft sciences," the problems are mostly defined by the readily observable symptoms, with basic data often unavailable and causal relationships largely unknown.[5] The few examples just discussed illustrate the large number of probable interactions; it is probably fair to surmise that many others are unsuspected at the present, but may turn out to be decisive at some future time.   Compounding the difficulty is the lack of broadly accepted decision criteria; many value judgments related to essentially incommensurable desires of human beings are again not the adequate diet for operations analysts.[6]

If methodology is found wanting, the availability of broadly trained generalists with the requisite combination of talent, wisdom, experience, and (most importantly) a sense of humility presents a problem just as serious, even if we assume that the proper motivating factors are present. Government organizations at the federal and state levels, mostly for traditional reasons, favor the functional (departmental) approach. Quite naturally, this attracts, motivates, and develops the functional "expert" with loyalties to a specific discipline or government endeavor; it explains why

---

5   (Note added in proof.) In not a few instances, the same comments could be applied to "hard sciences" as well. The current controversy about earthquake prediction should serve as a useful illustration and reminder.[55]

6   It would be difficult to improve upon Ashby's conclusions[56] which I cannot resist from quoting verbatim:

"All attempts to rely on quantifications in such decisions as these, to create them out of computer scenarios, to deduce them from cost-benefit balance sheets, are likely to make the decisions worse, not better; for in the process of getting hard data, the fragile values, the unquantified information, the emotive elements which nourish the public conscience, all run through the filter and are lost, and so the quantified information assumes an importance out of proportion to its real value."

Government activity is so often bogged down in the thicket of interdepartmental agencies and task forces; none having the appropriate staff work performed in time, of the requisite scope and level of thoroughness. Private corporations, except (perhaps) the more enlightened multinationals, also tend to do their strategic thinking by using their top managment people, carefully selected for single-minded aggressive devotion to specific product lines.   Often strategy planning staffs are used as mascots (in case the Board of Directors should express curiosity about the longer term future) but only seldom listened to when deciding about alternate or trail-blazing investments.[7] In the mid-1970's it was fashionable to suggest that the U.S. corporate strategic planning could, and perhaps should, serve as a model for national or supranational policy development. While there are many superficial similarities between the corporate and the geopolitical problems, these are not sufficient to inspire confidence in possible approaches derived from this analogy.[8]

---

[7]   Curiously, the most powerful advocates of long-term strategic thinking within many corporations are in the immediate vicinity of the Chief Executive; senior vice-presidents no longer burdened by the day-to-day operational responsibilities. They travel extensively, interact with senior people at home and abroad, they are an absolutely irreplaceable asset—if allowed to remain part of the central decision process.

[8]   There are indeed some essential differences in this respect between corporations and nations. First, corporations are, on the whole, run on an authoritarian basis. If an employee disagrees with the boss, being fired is a real possibility. No such threat of "liquidation" is being subsumed against any citizen in a Western-style democracy. Second, corporations have to abide by certain legal, regulatory, or institutional rules, valid for all competitors; the so-called "cut-throat" competition is usually accepted only in the metaphoric sense. Thirdly, the penalty for failure is not very serious, especially for large corporations. Profitable divisions are usually acquired by others; employees are retained with little loss of seniority or prestige. Even the chief executives clearly responsible for the demise of a given corporate entity have little difficulty finding other posts, unless gross dishonesty or misfeasance are clearly evident.

The biggest argument against the use of corporate techniques for national strategic planning is that they do not work real well, or at least not in adverse circumstances. Thus the electronic giants (General Electric, Westinghouse, RCA, and Raytheon) have failed to predict integrated microcircuitry based on semiconductors and the electronic data processing compo-

Talent and motivation are certainly available in the traditional refuges for generalists—think tanks, academe, and some policy-oriented foundations—but even these face contradictory pressures. Their "clients" or trustees are not, in general, persuaded that broad-gage, unfocused thinking is efficient; sometimes they are infected by the periodically fashionable "mission orientation," "end results," or "interest to the man-in-the-street" syndromes. When they are properly prepared for the respect of "soft" sciences, they tend to distrust the contribution of hardware, facilities and resources; the converse is also true when hard science and engineering specialists try to dominate the unrealistic pseudo-theoreticians dealing with the far more elusive social issues. Overshadowing it all, and related to the specific goal orientation, is the often observed tendency to "manage" the thinking phases of any large-scale project. Perhaps the real answer is in what Thomas[48] advocates for research in bio-sciences: Provide the "atmosphere" for broad-gage thinking, secure the participation of outstanding people of widely different backgrounds, then accept exploratory efforts based on hunches or intuition.

*

This chapter should be construed as a warning about the complexity of interactions among the elements of the environmental shifts and the corresponding (apparently desirable) policy changes that should constitute the adaptive responses of human societies. As such, selecting the proper and timely responses may turn out to be important or even vital to a particular group. This task is far from being simple; it is not certain that we have the tools and the people properly marshalled for a reasonable probability of success.

8 (Footnote continued)

nent market is dominated by corporations which did not even exist 30 years ago. Ironically, some of the companies cited in 1979 as using, and proposing the use of, strategic planning techniques for national policy development (General Motors and Ford Motor Company)[57] are among those which today are in serious difficulties and clamor most vigorously for Government protection against foreign competitors.

Whatever the methodology, it is well nigh impossible to make real progress except by giving full weight to the strongly interacting characteristics of superficially separate policy areas. Without any shadow of a doubt, the wide diversity of concerns must be approached as a *whole*; social, political, economic, and military matters are closely inter-woven and they can no more be separated for diagnosis and treatment than the parts of the living human mind and body, if these are to be understood and ministered to when given to illness.

It is probably wise to conclude that in the absence of empirical background, lacking the proper methodology to deal with not easily quantifiable value judgments and facing the difficulty of enlisting the required professional dedica-tion and competency, many large-scale "task force" or "project" type efforts are categorically doomed to failure. Those who suggest, no matter in what veiled form, that strategic planning at the national or international level can be competently accomplished by governmental authority, should be encouraged to reconsider their position in the light of the arguments presented here.

We must recognize that even with the suggested premises, the problems of policy formulation can not be solved by analysis. Solutions, if any, result eventualy from a *continu-ing* process that is essentially and unalterably *political*. Ideas and concepts, arising from serendipity or derived from reasoning deemed rational, are submitted to the scrutiny of the (theoretically) well-informed voters; if they carry reasonable consensus, they become enshrined eventually into public policy. This is, after all, the time-tested relatively peaceful technique for resolving conflicts in a democratic society, and we have seen in all our foregoing discussions to what great extent policy formulation and implementation are perfused with conflict situations.

Government officials of all three branches, rather than arrogating to themselves the full authority for policy formu-lation, would be well advised in focusing their attention on preserving the integrity of the policy development process.[58]   By no means is it an abdication of their responsibilities of leadership to articulate clearly the alternatives of public policy together with the relevant issues, but then to direct their staffs and to inspire the professionals at large to bring their collective efforts to bear on the development and dissemination of objective and relevant information within their respective areas of competency.

With these words of caution, we are now ready to face our conclusions.

Part 4
SUMMARY

CHAPTER XV
Conclusions

This is the final chapter of a rather lengthy story. We started together on a journey, enticed by the author's promises; having reached thus far, the weary reader is entitled to the reward of judging whether or not the promises have been fulfilled.

Examining the multiple causes of dissatisfaction and stress within the affluent and complex Western societies, we found that they can be attributed plausibly to *evolutionary misadaptation*, which is the failure to accommodate some major environmental changes likely to impinge upon their common fate. Even when perceived, the customary and characteristic decision processes of democracies are ill-suited to elicit the adaptive responses that may attenuate the growing incidence of stress and help perhaps to steer away from irreversible catastrophes.

Subjecting some of our prevailing geopolitical premises to critical scrutiny from the adaptiveness standpoint, we found that they may indeed benefit from some significant (although by no means radical) modifications. The premises under examination can be conveniently clustered into four groups: (1) National (domestic), which deals with the sociopolitical areas within sovereign boundaries; (2) International ("West"), which addresses the relationship among the industrialized nations; (3) Global-economic, involving the first two, but mostly in relation to the Third World; and (4) Military, which relates to the future role of force among nations.

In regard to the domestic goals, it is urged that the Western nations might do well in shifting some of their concerns to societal stability in addition, rather than in contrast, to their traditional devotion to enhancing the rights of the individual citizens. To strengthen the relationships among the industrialized nations, it is proposed that the time has come for some supranational entity to be formed, which would help them all, deliberately and progressively, in the ideological, economic, and military competition against the rest of the globe, at the cost of relatively trivial sacrifices of their individual sovereignty. With respect to the less-developed countries, instead of trusting some "laisser faire" type mechanism, the industrialized Western societies should evolve and sustain the long-term investment required to strengthen (and if necessary, to create for this specific purpose) the trade gradients which should serve as underpinning for a robust global economic structure. It should not, but it will, surprise many of us that the real "correlation of forces" in these matters favors the West rather than its competitors in Soviet Russia and elsewhere. If properly exploited, this simple, fundamental fact may prove decisive in the efforts to enlist the long-term allegiance of the less-developed countries to the Western camp.

All this has a nice optimistic ring to it. Assuming that all the arguments presented here are complete and fully justified (unfortunately a strong premonition should be always present to the effect that they are *not*), and assuming further that they can be explained to every individual on earth, to each in the appropriately intelligible language), even then world history would simply not evolve in the smooth "rational" way implied by the modified premises. People are well known to be "irrational" (according to other peoples' judgment), and we are harkening back to a legacy of many millenia for using threats and violence whenever our wants or needs (sublimated into policy objectives) are opposed by others. The stress-free future, so glibly implied in the first three premises will be endangered by those who are less fortunate in their possessions and by those who hold that the use of violent means are justified by the righteousness of their cause. For the West, failure to refurbish its military might is tantamount to surrendering the freedom they enjoy, and the ideals they cherish; far gone are the days when far distant populations could not wage war effectively across the oceans. Again, surprisingly, we find that, while many of the current Western military posture elements are weak or threatened by the Soviets, it is perfectly possible and within the collective means of the U.S. and of its allies to recover their military strength in time to *deter* any nuclear or conventional conflict, to *control* the pace of escalation in any shooting war and to *prevail*, if fighting becomes necessary. Not only is such military recovery possible and desirable, it is nothing less than vital. The purpose of military power for the West is not to dominate over a fully subjugated planet, with its populations exploited and resources plundered; it is the *preservation of a relatively conflict-free path*, over the next 25 to 30 years, which allows the now less-developed nations to be integrated into a modern global society. This is what is seen as a relatively attractive rationale for Western military power. If the currently formidable military machine of the Soviets can be held in check for the next few decades, there is at least a fair chance that they will acquire more vested interest in the peaceful status quo than in pursuing a policy of violent confrontation. This may owe as much to the deterrence by Western forces as to the evolution and increased stratification within their own society.

*

When it comes to policy formulation and implementation, another pleasant source of amazement is in store for us: Hardly any of the suggested policy objectives are radically new or require fundamental revision of our past values or thought patterns. None of them is found in contradiction with the Constitution as it has been amended and applied over the past two centuries; the changes are more a matter of focus and emphasis to accommodate the realities of modern technology-saturated societies. A fundamental question is raised about the role of responsible and informed citizens, the purported true guarantors of the integrity of our political processes, when information is being generated in massive quantities and without quality control, and is being permitted to pervade their conscious thought processes and even their subconscious. Professionalism among those responsible for mass communications is being advocated as the means for striking a balance between arbitrary censorship and unbridled freedom.

Government-business relationships within the U.S. are seen as deserving serious attention and as requiring changes in the direction of true partnership. This is the case to a large extent in Japan and in the Federal Republic of Germany; it is gaining ground in France as well. We can no longer afford the luxury of doctrinal dissentions of this subject; there is an urgent task of marshalling all the intellectual and physical resources of the nation. Goal setting, with the traditional checks and balances is the proper role of the Government, monitoring progress and law enforcement are the necessary adjuncts, but not competition for any significant share of the economic activity. The latter has proved to be in much better shape when left to the true and tried profit motive of private enterprise.

When public wealth is being charitably transferred to the less fortunate,[1] it is found that all mechanisms that support beneficiaries without requiring countervalue in work or at least in effort toward self improvement are socially undesirable in the long run. There should be strong incentives established (and seen that opportunities be available to all) to participate actively in the productive process.

Two major investment areas, currently neglected, are seen to contribute potentially to the successful implementation of the suggested policy objectives. First, as of now, the intricately interconnected and exquisitely fragile industrial societies are at the mercy of violence, natural disasters or manmade catastrophies. A substantial capital investment and a significant redirection of our labor force would be required in order to provide even a minimal protection level against the effects of nuclear warfare. The strategic interests of the West suggest that such investment may be

---

[1]   (Note added in proof.) The currently fashionable term is "truly needy" (March 1981) in sharp contrast to previous Administrations, when the proper designation was "underprivileged." It should be obvious that this change is more than just vocabulary.

necessary, and it is observed that there are common elements between the improvements in societal resilience and other desirable economic endeavors. The second major investment is rendered necessary by the finding that trade gradients ("what can we sell to pay for what we want to buy" and *vice versa*) must be continually enhanced and strengthened if they are to serve as support for a viable global economic framework, involving the less-developed countries in particular.

A different type of sacrifice is needed to bring forth serious progress toward the gradual aggregation of the Western nations in the economic, military, and (eventually) the political domain. There are strong and compelling reasons for the U.S. to exert leadership, but this should not mean, in anybody's thinking, the equivalent of domination. In terms of cultural background, human and natural resources, productivity, and per capita wealth, our Western partners are fully comparable to the U.S.; the initiatives, responsibilities, burdens and future blessings should be shared to the fullest.

To recover the appropriate level of military preparedness, the West must, of course, increase the effective level of military expenditures in order to compensate for several past decades of complacency and overemphasis on "reordering our national priorities." Socially motivated expenditures and capital investments are, of necessity, constrained by the increases of military budgets. The military establishment will have to streamline its acquisition processes in order to make them more cost-efficient and to provide superior equipment to the deployed forces at the proper time and in sufficient quantities to ensure superior combat power. Beyond the procedural improvements, we have discussed a few general principles in weapon system development that may further improve the resulting force structure for a given expenditure level.

All the premises and derived implementation aspects have one major common element: they place new limitations and constraints on individual citizens. To those who question the need or the urgency for accepting such constraints, a simple one-paragraph answer distills the essential:

Irrespective of the reasons, the past rights and wrongs, we (humans on this planet) have allowed our breed to multiply, without spreading at the same time the benefits of education, discipline, and shared value judgments. On the other hand, we have developed and allowed to spread the most dangerous technologies of propaganda and weaponry. These conjoint elements create pressures on the advanced societies, which in turn are being transmitted as constraints affecting their individual citizens. The relatively mild constraints and sacrifices implied here are thought to be most acceptable in comparison to those which would result from large-scale wars and other violent convulsions.

*

Several times the opportunity presented itself to praise the wisdom or "visceral prescience" of citizenry reared in democratic traditions. More than once it was observed that many of the plausible or desirable implementation moves are in fact taking place, slowly perhaps, but nonetheless perceptibly. If so, the question arises about the implicit urgency of the recommendations. Why do we have to do anything? What is so unique about the present, or the forthcoming decades? These are indeed valid questions and probably deserve better answers than the preliminary views expressed here.

In regard to internal domestic matters, we are concerned mostly with slowly evolving trends and attitudes; the important factor is the intellectual and emotional commitment of the policy makers and of those who shape public opinion. Unless the citizens of this nation recover the will and the stamina to act resolutely in their own long-term interest, risking their fortunes without sacrificing their principles, the nation will not be able to support morally and materially the strength required to make its voice respected in the painfully protracted evolution toward a more secure and stable world. This is why progress toward a *truly* just society, which in the meantime will have retained its cohesion and productive efficiency, is so important—people are more than willing to work hard and even to incur serious risks or discomfort, if they can envision realistically a more promising future . . . especially if it is *their own* future.

The progress toward the Western "commonwealth" is to a great extent also a matter of resolve and attitudes; the specific timetable for achieving diplomatic and legal milestones is probably of secondary importance. This is so precisely because the Western nations are perfectly capable of acting pragmatically in concert when they perceive the congruency of their interests. The problems are more a matter of differential perceptions and of persuading the nationally elected leaders to court disaster at home by supporting some worthwhile but remotely distant international goal. The real urgency is in defining and establishing the essential cooperative mechanisms that allow them to act together in emergencies, which are only too likely to occur within the very near future. Upheaval in Europe, possibly arising from unrest within the Soviet military alliance (ironically termed the Warsaw Pact[2]) may precipitate East-West confrontation; the volatile Middle East, possibly further destabilized by Soviet moves, is certainly not a reliable source of the oil flow still vital to Western Europe and to Japan, and may call soon for unambiguous joint Western action. These two sources of possible major perturbations (to put it mildly) are upon us right now: whatever accelerates the aggregation of the Western nations, whatever contributes to their ability to act in concert, should be encouraged and supported.

The development of the global economic framework is mostly important in the 10 to 15 year perspective. By the end of this century, the very large majority of the world's population will live in what today are the less-developed nations. This immense mass of humanity will seek, nay demand, what (according to our own paramount beliefs) is their birthright: life, liberty, and the pursuit of happiness. In more mundane terms, they will strive for food, shelter, health, progeny, security, and a voice in deciding their own destiny (not necessarily in this order); the West will have to deal with their striving. As of now, the terms of transaction are still largely open; we may trade goods or services, or again we may, to put it bluntly, trade missiles and warheads. This is, admittedly, generalization and oversimplification,

---

2   Parenthetical clause added in proof.

but gives some weight to the suggestion that the efforts to renew the vigor, and to protect the stability, of world trade are by no means premature; in point of fact, they may well be overdue by the time the results could be felt.

Progress in the interconnected domains of domestic, international, and economic affairs may, or may not, be matters of high and compelling urgency, but there is the fourth domain: that of military preparedness, seen by some as the dark horse of the Apocalypse. All the potential, all the promise of progress in the other areas, will be utterly frustrated or destroyed should the Soviet Union gain acknowledged military supremacy over the West—there are competent judges of military matters in both camps who clearly believe that this is in the process of happening right now. In the nuclear age, intercontinental distance no longer shields the peace-loving nations, neither will the rapid rate of onset of plausible war scenarios offer the luxury of mobilization and the hope of eventual recovery of a superior military posture. Nuclear weapons are part of our world now: we have seen that they closely interact with, and are inseparable from, other aspects of military power. We all hope and pray that they be restrained by negotiated arms control treaties, but this has certainly not happened to date and certainly not for lack of trying by the West. We are *now* in a state of precarious balance and the momentum strongly favors the Soviets. This momentum must be urgently reversed; all the other components of progress are contingent upon the implications of this simple, stark statement.

*

Perforce, a few important topics have been only glimpsed at; one should hope that briefly mentioning them would prompt further inquiry.

Partly based on the ominous Soviet military build-up, but mostly because of what appears to be single-minded persistent pursuit of a long-term strategy clearly lethal (if successful) to the West, we have given preponderant emphasis to the ills which affect the Western societies. Even a superficial glimpse at the opposite camp shows that the same environmental causes impinge upon their present and future, perhaps to a much greater extent than suspected. It is possible, and perhaps probable, that these impacts would produce profound and relatively rapid changes, not all necessarily unfavorable to the West. As technology is spreading even within totalitarian societies, they too have great difficulties in controlling the information flow which gains access to their population. Their mastery in propaganda and indoctrination will do them little good once the ubiquitous (space) satellites give the tiniest transistor package access to Western communications, conveying truth or propaganda or a mixture of both. As more information is demanded and gained by their citizens and by those of their reluctant (political) satellites, divergency would be followed by disaffection and dissidence. The minority population growth rates exceed by far that of the Russian core of the Soviet Union; their future orthodoxy may be overshadowed by renascent racial and religious allegiances. The Soviet government may have to recognize the finite limits of their natural resources; their agricultural and industrial base may be found grossly insufficient to support their world-wide political ambitions. Should the West elect to modernize its military strength with competency and dispatch, the Soviet Union may well find that their enormous sacrifices, patiently accepted for the past 30 years or so, will have been to no avail. None of this would contribute to the emergence of a stable, repression-free society with beloved and trusted leaders; none of it can be held up for long as the workers' paradise to the emerging nations as an example to follow. In fact, in no country has communism been freely accepted or perpetuated (and this statement covers specifically Soviet Russia) except through the force of arms—not exactly a ringing historical endorsement for any political system.

The major alternatives to the scenario implicit in our discussions have not been seriously explored. For instance, one could have conjectured a future where Japan, pressured by the lack of resources[4] is abruptly or gradually detached

---

4 The 1939-1940 oil embargo imposed upon Japan, which, according to many historians, led to Pearl Harbor, may well be remembered in this context as a precedent.

from the Western community, and leads the immense Asian human reservoir (China, India, Indo-China and Indonesia) toward rapid evolution and eventual superpower status. The timetable for all this to happen is usually left unspecified, but can be hardly less than a generation. I hold, as a matter of largely unsubstantiated personal opinion, that such a scenario is implausible, but I am more than willing to concede that it may benefit from further exploration. It might be added that no suggestion of this type has arisen in discussions with investigators of oriental extraction.[5]

*

Before taking leave of my patient reader, I would like to sound a note of optimism. A clear apperception of what has happened should greatly help to avoid large-scale future mistakes, and this effort should be seen as an attempt to enhance our perception. I hope that at least some of it will deserve the lasting interest of those motivated to validate, to expand, or to build upon the construct outlined here. Many details are admittedly sketchy or entirely missing, but I hope that the basic message will not be missed. We have a reservoir of background knowledge and some of the intellectual resources required to approach the problem of our own societal evolution dispassionately and competently. Policy alternatives can be explored, and if found attractive, submitted to adjudication by the citizenry and their representatives. There is hope, for the first time in history, to do this on a global basis, with a reasonable hope that actions can be taken before the occurrence of large-scale convulsions or irreversible catastrophes. Other things being equal, this type of effort should attenuate the many turbulent stresses which otherwise accumulate to distress our present and to darken our future. No professional challenges can be intellectually more stimulating; none of them should be potentially more rewarding in view of contemplating the momentous consequences of failure.

[5] Cf. Footnote 9, Chapter XI.

To the last I have left the special role and responsibility of the United States. This nation has been blessed with ample land, beneficial climate, and immense treasure undergound. In comparison to their old homeland, immigrants found this land to be limitless and incredibly bountiful. It offered to all the opportunity for self-reliant hard work and favored the true blossoming of democracy. It should be seen as more than a coincidence that the U.S. is the first nation that evolved successfully a multi-layered structure of sovereign governments, capable of handling the problems of an immensely variegated country, inhabited by a strongly pluralistic society. There is grandeur and beauty in the thought that the U.S., having benefited from this special dispensation, should shoulder the responsibility for sharing its political genius with the rest of mankind.

* * *

Appendix Notes

– NOTE A –

ADAPTIVENESS AS A SOCIOPOLITICAL CONCEPT

When speaking of adaptiveness in reference to an organization as complex as a whole nation, one should think about the large-scale group behavior of its population, not the details of individual behavior. People will act as a group either as the result of constraints imposed by formal laws, treaties, or (for smaller groups) contracts; or such action will be empirically observed as a "normal mode" with relatively small variances. Thus, the nation as a whole may exhibit wholesome characteristics of vigor, while some individuals or small groups suffer stagnation and decay typical of maladjustment, or *vice versa*.

In the context of adaptation being a measure of efficiency in utilizing resources, the term resources is used to designate whatever can be used to the advantage of the nation. It covers physical resources, such as soil, water, mineral wealth, and climate; it includes human resources such as intelligence, experience, skills, work habits, and competitive instincts. It obviously includes synergistic combinations of physical and human resources; these can be in various degrees available and/or developed within the known economic and technical limitations.

Adaptation is a concept always defined in relation to a given environment. The term environment is used in a much broader sense than that which is customarily accepted. It retains its etymological meaning, i.e., factors derived from the surroundings, or elements external to the group. As such, they may be constant or immutable and thus be independent of the desires or the actions of the group. They may be "external" in the sense that they are mostly independent, but in fact they are subjected to more or less rapid changes by natural causes or by the inadvertent or deliberate action of the group itself. Still some other "external" environmental factors are mostly (but not entirely) under the conscious or inadvertent control of alien groups, possibly hostile to our own. These again may, or may not, be susceptible to modifications through our own actions.

It is expedient, for our purposes, to extend the concept of environmental factors to include those which are "internal" to the group. Stated otherwise, elements of the internal environment are closely (but not exclusively) associated with the nature, characteristics, and operation of the group; they include specifically the physiological, intellectual, and behavioral characteristics of the population. They, in fact, share all of the attributes specified for the "external" environment, i.e., some of the elements may be essentially immutable, such as the genetic inheritance of a given population; some others may be subjected to more or less rapid changes both through inadvertence or through deliberate action; still some others may be subject to influence, or even hostile manipulation, by actions of external groups. It can be argued with some measure of justification that the "internal environment" is probably more important by far than is generally suspected in determining the degree of adaptation of a nation, or for that matter, its "adaptiveness" (capability for adaptation).

It should be recognized that the differences between external and internal environment factors are nonessential matters of degree, rather than of kind. We may generalize by suggesting that the difference between "immutable," slowly changing or rapidly changing environmental factors, is also amenable to definition as a *continuum*, and so is (not surprisingly) the artificial dichotomy between deliberate and inadvertent changes. All "deliberate" human actions may have impacts of often unintended and unsuspectedly serious consequences, the reciprocal not being true by definition. The separation between hostile and competitive is also a matter of perspective; action by an external group may be perceived as healthy competition or diabolical machination, dependent on the analyst's vantage point.

Even the term so innocuous as "beneficial" may need some elaboration. Who has the wisdom to determine that a given development is beneficial? Who are to be the beneficiaries, one or many diverse segments of the population? Is it obvious that some benefits are so general that the majority of the population should favor the change? Are some subgroups capable of appreciating the social value of a given change?   Are the (obvious) beneficiaries justified in trying to change the attitudes of those who would tend to object? If so, what are the permissible means?   If violence is out, are threats permissible?   How about drugs or intoxicating beverages?   The list of questions can go on, and unfortunately, there are no good answers except the truth: We do not really know.[1]

− NOTE B −

ROLE OF NONVERBAL COMMUNICATIONS

The role of nonverbal communications in human interactions is well established.[59,60]   The basic objective of communications is the perception, by the recipient, of the sender's purpose when transmitting a message.   Both conscious and subconscious elements are involved in the perception mechanism, which consists essentially in the more or less accurate match (association) of the incoming message pattern against those stored in the recipient's memory. In higher primates and in humans, the brain is the primary site of conscious memory, but many biologically important patterns are stored ("imprinted") without the individual's *conscious* awareness or ability to recall.

---

[1]   In absence of formal or logical background, perhaps a working definition should suffice. Each societal group, including the nation as a whole, develops mechanisms for arriving at decisions affecting the group and for enforcing the consequences of these decisions. These roughly correspond to the legislative and executive functions of government; sometimes, but not always, they are balanced by the judiciary function. The term "beneficial" could be operationally defined as being seen as such by the decision makers of the group. If their judgment is right, the group will eventually benefit; if it turns out to be wrong, the group will suffer; if judgments are consistently wrong in the long-term average, the group will disappear under the pressures of competition, unless it finds in time ways to change its decision makers. Long-term survival (implying also maintenance of relative levels of prosperity and growth opportunities) is thus used as an operationally "objective" criterion for being beneficial. By using this working definition, the ability to decide whether or not a given development is beneficial becomes an element of adaptiveness of the same type as the ability to perceive, in time, modifications in the environment.

It is difficult to doubt that, given sufficient number of information bits,[2] any message can be transmitted by means of relatively simple patterns. Dependent on the degree of congruence of these patterns with the training, memory, experience, and imprinting of the recipient, these message patterns will be more or less effectively translated and understood. Thus, warning signals, traffic lights, Morse code, Braille alphabet, and playing cards are relatively simple examples; printed words, commercial logograms, comic strips, mathematical symbols call for increasingly sophisticated memory and pattern recognition capability. Graphological interpretation of handwriting and communications inherent in the appreciation of abstract art require considerable professional experience and erudition. Together, conscious memory and subconscious imprinting enable individuals to extract messages from multisensory information flow (and to interpret them for individual or social benefit) with truly astounding speed and reliability, certainly far beyond any conceivable future state of the art in automatic data transmission and processing.[3]

Even though apparently far-fetched, these observations are of considerable relevance to the main theme of our inquiry. In the past few thousand years mankind has considerably evolved in the cultural (as contrasted to the biological) sense; in terms of social organization the definition of common purposes has progressed from the tribal to the regional to the national level, and in many cases important political bonds operate at the level of federations encompassing whole continents. Human social destiny is

---

2   Without attempting rigor, this would be in engineering terms the *rate* of information transmission (itself proportional to communication bandwidth) multiplied by the message *duration*.

3   As an illustration, close and protracted proximity of young adults of the opposite sex appears to produce, even in the dark, multisensory stimuli quite appropriate for the biological and social purposes at hand. Advanced degrees in biochemistry, physiology and psychology do not appear as necessary prerequisites. By way of contrast, transmission of the same messages over artificially structured message patterns, such as printed stereochemical models of the pheromones operating in such situations, would be considerably less effective, even between otherwise qualified partners highly trained in these disciplines.

thus directly influenced by distant groups which may act through messages, economic competition or cooperation, and through military threats or alliances. In the last hundred years or so, for all practical purposes, the information flow beyond the immediate family, tribe, or neighborhood has become electrical or electromagnetic in nature, and this transformation is all but complete as of the last two decades. "Electronic" communications are complacently described as "broadband" and indeed in comparison with the original telegraph, telephone, and even radio broadcast, the available communication links offer bandwidths by several orders of magnitude beyond those available as little as twenty years ago. Alas, for reasons of efficiency in the distribution, the *available* bandwidth is promptly subdivided into many individual channels, each of them just barely adequate for their assigned role. Even color television, with all its impressive performance and potential, is pitifully inadequate in comparison to the rich, high-resolution, multisensory information flow available through direct, person-to-person contacts.

Thus, while humans are interdependent on a large scale (i.e., individual lives are influenced by actions extraneous to the local political groupings), communications among the interacting groups are only superficially adequate. This is the reason why people are still instinctively attracted to meetings where political celebrities are present "in the flesh"; this is why live performance of artists is still rated far above their electronic presence; this is why people still spend a considerable proportion of their leisure budgets on travel in distant foreign parts where different cultures may be explored first hand. Even such unemotional and cost-conscious groups as businessmen, scientists, and other professionals seem to hold the view that personal contacts and professional meetings such as seminars, workshops, and symposia rate far above the messages obtainable through correspondence, teleconferences, or published proceedings—even with the stern eye of the IRS watching over the extracurricular activities.

Even if and when ultra-wideband communications using optical transmission links are perfected over long distances, this fundamental need to meet fellow humans in person by direct contact may remain the major driving force sustaining the commercial air travel market. The opportunity for direct personal contacts may be the principal tool whereby deliberately truncated or distorted messages can be corrected and countered.

— NOTE C —

ANATOMY OF A CRISIS - THE DC-10 ACCIDENT

Whenever a significant event comes to disturb the routine of society, the news media immediately concentrate their considerable resources on providing prompt, complete, and indepth coverage in order to keep their audience and readers informed and (incidentally) to "sell" their services. Professional animus and downright competition conspire to enhance the value of promptness (sometimes at the expense of reliability or even of plain truth) especially when the news has direct, immediate relationship to public concerns or, even more so, when the implications shed favorable light on the causes supported by the media.12

It is instructive to review a recent example by following the "official" statements published in reputable newspapers, as summarized in the New York Times Information Bank.61 When attempting to assess its probable impact on the public, one should keep in mind that, in addition, the public was subjected at the same time to a saturation barrage of several months' duration, by television, radio, and news magazines on the same subject. Here, then, is a much abbreviated summary:

- NYT [4] 5/26/79:   All 272 aboard killed...American Airlines DC-10 Chicago to Los Angeles...worst disaster in U.S. aviation history...

- NYT 5/27/79:   G.N., who gave up his seat on doomed American Airlines DC-10 to woman who pleaded for it, describes feelings...

[4] New York Times

- NYT 5/27/79: NTSB probe of American Airlines DC-10 crash centers on loss of engine and almost simultaneous cutoff of cockpit recorder...

- NYT 5/28/79: NTSB says breakage of 3-inch bolt because of metal fatigue evidently caused engine and mounting to fall off American Airlines DC-10, Chicago; Board urges emergency inspection of all "attach points" where engine mountings are fixed to wings on DC-10's...

- NYT 5/28/79: FAA orders airlines flying DC-10 jetliners to immediately inspect engine-mount bolts...engine-mounting rear bulkhead, found to have abnormal markings...flight recorder indicates that pilot's inability to keep craft under control was caused by progressive loss of hydraulic fluid...

- NYT 5/30/79: FAA grounds all 135 DC-10 jetliners owned by US airlines, citing "grave and potentially dangerous deficiencies"...indicates that structural problems in engine mounts appear to be more widespread and serious than first suspected... McDonnell Douglas and NTSB concur in grounding order...

Up to this point, the main concerns reflected by the media are the causes of the accident and the means for enhancing air travel safety. Now, the attention begins to shift to the "business implications":

- J of C[5] 5/30/79: Risk Research Group chairman James Bannister expresses concern over current aviation insurance rates...

- NYT 5/31/79: McDonnell Douglas Corp confident that it will continue to sell DC-10 jetliners despite recent crash of American Airlines DC-10, Chicago...

[5] Journal of Commerce

- NYT 5/31/79: Many of US domestic airlines 138 DC-10's resume flying...FAA spokesman says inspections "turned up more deficiencies than found in previous inspections"...

- Wash. Post 5/31/79: US Aviation Insurance Group president estimates that recent crash of American Airlines DC-10 could result in payments totalling $100 million...

- NYT 6/1/79: FAA reports problems in engine mounts of 37 of 137 DC-10 jetliners operated by US airlines...

- NYT 6/1/79: Mother of Schenectady man killed in American Airlines crash, Chicago, sues American and McDonnell Douglas for $1 billion...

- Wash. Post 6/4/79: Prospective damage suits stemming from recent American Airlines DC-10 crash are expected to generate strong competition for cases among nation's small and exclusive group of "air disaster" lawyers...

- NYT 6/5/79: NTSB urges FAA to advise airlines that maintenance procedure of raising and lowering pylon with engine still attached may have led to American Airlines DC-10 crash...

- WSJ[6] 6/5/79: Passenger aversion to DC-10 following American Airlines crash in Chicago is causing some to seek different planes...

- NYT 6/7/79: FAA head Langhorne M. Bond orders immediate, indefinite grounding and suspension of design certification of all 138 DC-10 jetliners operated by US carriers...says discovery of flaws in still more craft strengthens fear of design defect...

[6] Wall Street Journal

- NYT 6/8/79: Probe of DC-10's focuses increasingly on craft's hydraulics systems...industry study of DC-10's indicates 89 cases of hydraulics problems over 16-month period ending in December '73... Administration ordered major modifications of systems; system's failures and speculation on system failure as factor in American crash discussed; Federal District Judge Aubrey Robinson alters his temporary restraining order grounding all DC-10's by giving permission for craft to be flown on "ferry" (non-passenger) flights to maintenance bases or on test flights to determine craft defects; DC-10 operators, Western Airlines, Northwest Airlines and World Airways, ask Robinson to make them parties, with Airline Passengers Association, to case involving...McDonnell Douglas informs Safety Board that it will not seek reversal of grounding order...

- WSJ 6/8/79: Domestic and foreign airlines are losing millions of dollars in revenue during peak summer travel season because of grounding of McDonnell Douglas DC-10's...

- WSJ 6/8/79: WSJ editorial commends FAA's decision to ground DC-10, but hopes plane will not fall victim to antitechnology hysteria...

- Wash. Post 6/10/79: Handling of DC-10's after crash of American Airlines DC-10 in Chicago damaged FAA's reputation;...will have to reexamine its testing, certification and inspection procedures.

- SFC [7] 6/10/79: Air crash litigation is proving lucrative for attorneys who file wrongful death suits for next-of-kin of crash victims; air crash specialists' fees are usually 25% of settlement...

---

[7] San Francisco Chronicle

- NYT 6/20/79: Thirteen major European airlines clear most of their 58 DC-10's grounded since May 25 crash of American Airlines craft and subsequent discovery of structural flaws...

Roughly a month after the accident, the political and economic consequences are increasingly in the forefront of new coverage:

- SFC 6/24/79: House Government Activities and Transportation Subcommittee chairman John Burton has been criticized for his "ungentlemanly" treatment of FAA Administrator Langhorne Bond during hearings on American Airlines DC-10 Chicago crash...

- NYT 7/2/79: American Airlines employees purchase full-page advertisement...to rebut allegations by Federal Aviation Administration that May 25 crash in Chicago was caused by faulty maintenance ...

- NYT 7/9/79 NYT: FAA expected to announce within day or two lifting of ban against DC-10 jetliners...

- NYT 7/11/79: FAA report blames in large part American Airlines maintenance procedures for May 25 crash of DC-10 jetliner in Chicago which killed 273 persons...

- NYT 7/13/79: Bill Moyers' comment on American Airlines DC-10 crash...noting overly comfortable relationships between aircraft manufacturers and Federal Aviation Administration ...

- NYT 7/14/79: FAA head Langhorne M Bond lifts ban that grounded nation's 138 DC-10's...

- Fortune 7/16/79: Investigation surrounding crash of American Airlines DC-10 flight 191...May 15, '79 killing all 271 aboard suggests that maintenance procedures, aircraft design and FAA regulatory lapses all contributed to worst air disaster in US history...

- Editor and Publisher 8/4/79: Corporate gadfly Lewis D. Gilbert charges...that stockholders were raising questions about American Airlines' maintenance practices at meeting eight days before DC-10 crash in Chicago...

- NYT 8/9/79: American Airlines production manager William S. Fey...admits American used method of removing engines for maintenance that was not approved by McDonnell Douglas.

- SFC 8/11/79: American Airlines' insurance payment on DC-10 plane that crashed at O'Hare International Airport (Chicago) on May 25 '79 caused '79 2nd quarter to be American's most profitable quarter ever...

- NYT 8/12/79: Some airline passengers continue to shun DC-10's as result of American Airlines May 25 crash...

- Aviation Week & Space Technology 8/13/79: National Transportation Safety Board officials are expected to mandate more stringent quality control in commercial aircraft manufacturing and maintenance...

- NYT 8/19/79: McDonnell Douglas Corp delays placing newspaper advertisements designed to restore public confidence in DC-10's following American Airlines crash in Chicago on May 25...

- Chicago Tribune 8/26/79: Discovery by United Airlines mechanics last May 29 of cracked engine pylon on DC-10 may have averted crash like that which killed all 273 passengers aboard American Air Lines DC-10 on May 25...

- Time 8/27/79:   Some families of victims of American Airlines May '79 DC-10 crash are seeking damages from American, McDonnell Douglas and their insurers through aviation accident lawyers...

- NYT 9/16/79:   American Airlines...says it is committed to "extensive structural and system changes on its thirty DC-10's...that it has asked McDonnell Douglas Corp...to make list of modifications and that McDonnell has agreed to make them...

- NYT 11/17/79: FAA announced American Airlines paid record $500,000 fine for improper maintenance of DC-10 airplanes...

- NYT 12/21/79: NTSB says blame for crash of American Airlines DC-10 in Chicago in May should be shared by airline, McDonnell Douglas Corp and FAA...

- NYT 12/22/79: NTSB reports that combination of improper maintenance and vulnerable design of key structures led to May '79 crash of DC-10 jumbo jet in Chicago...

The net result is an improved set of FAA-industry regulations covering certification and monitoring of maintenance procedures, a more stringent enforcement of the manufacturers' maintenance recommendations, and (perhaps) some basis for settlement of damage claims. Whether or not the sensational headlines were necessary or useful contributions is left to the readers' appreciation.

— NOTE D —

GROWING PROPORTION OF MARGINAL WORKERS

Ever since the advent of industrialization, the economic role of those not fortunate enough to own land, other forms of wealth, or highly marketable skills has been the subject of sharp controversy and the cause of many upheavals. At the beginning of the 19th century, it was still perfectly respectable among reputable economists to suggest that starvation is the appropriate mechanism for regulating the availability of unskilled labor.[8] Somewhat closer to our time, the rise of labor unions in many cases mitigated the harshness implicit in this type of regulation, but coupled with the politically desirable pressures for minimum-wage laws, it also contributed to the growth of the marginal producer class, i.e., those whose labor is not accepted by the "market" at the prevailing prices. Among the several factors setting these prices, the economic value or usefulness of the work to be performed is the most obvious. The value of the workers' output may be less than the market price for any combination of reasons such as physical or mental handicaps, insufficient education, training, or motivation. In addition, legally determined minimum thresholds, the expectations of workers, and the competition from socially motivated transfer payments (welfare, unemployment, training, etc.) increase the effective price at which people are willing to "sell" their labor. Since starvation or other unpleasant penalties are no longer acceptable as motivators and since subjective penalties, such as loss of social standing or disapproval of peer groups, are no longer as powerful in today's large anonymous communities as they used to be in the past, the number of individuals entering the "marginal" group rises inexorably.

---

8   "When the market price for labour is below its natural price, the condition of labourers is most wretched; then poverty deprives them of those comforts which custom renders absolute necessaries. It is only after their privations have reduced their number, or the demand for labour has increased, that the market price of labour will rise to its natural price, and that the labourer will have the moderate comforts which the natural rate of wages will afford." (David Ricardo, 1817, as quoted by Hardin.[62])

The skilled blue-collar workers, the lower and middle segments of the traditional "white collar" class should not delude themselves in regard to "marginality" being intrinsic to the lower rungs of the economic ladder. In fact, many professionals and para-professionals traditionally associated with the middle class sooner or later face the same problem for a very simple reason: As long as the production mechanisms [9] are determined by the profit motive and the competitive climate, any technical improvement beneficial on narrowly construed economic grounds will be introduced as soon as the state of the art makes it available. Since the pace of technology research and its applications to industry have much accelerated in the past few decades,[10] and since the innate adaptiveness of individuals (their capacity to learn new intellectual or manual skills) has little changed in the same time period, increasingly larger population segments find themselves in the midst of the process of becoming marginal. In more general terms, in many sectors of economic activity the skill spectrum required in order to function effectively in a post-industrial society can be observed to shift in the direction of higher skill levels. The pace of this shift appears to be faster than that which would correspond to the adaptation capacity of the affected workers; those who fail to adapt sooner or later become "marginal," even though protected in the near term by unions, seniority rules, and other similar devices. Most significantly, instead of affecting the unskilled or those handicapped by physical, mental or social conditions, the threat of becoming "marginal" through obsolescence is present for a growing fraction of skilled and even professional workers.

If all this sounds theoretical or far distant, let us consider a few specific instances: (1) Insurance policies and premium notices are now automatically processed; many clerical jobs are being eliminated, while the demand for software

---

9   This term is used in a rather broad sense. It includes manufacturing, but also all the planning, design, organization, and administration of the associated business activity; it includes marketing, selling, service industries, in fact any activity pursued in view of profit or economic advantage.

10   In fact even more so in Europe and in Japan than in the U.S.

specialists and equipment maintenance personnel increases. (2) Computerized travel reservations and routing much reduce the skill and the experience necessary for the agents at the airline counters.   (3) The general availability of magnetic tape recorders has just about eliminated the need for stenographic skills in legislative proceedings; the same technical threat hangs over the traditional court recorders. (In the business world many higher level executives still prefer dictation to live secretaries, probably on the grounds that the instantaneous feedback increases cogency and reduces verbosity.) (4) The advent of microcircuitry and computer-aided diagnosis has essentially modified the skill requirements for electronic appliance repair and for automobile maintenance; the use of automatic diagnosis and laboratory tests for medical purposes has in turn brought experts in electronic data processing into clinics and hospitals, with the corresponding relative decrease in the number of skilled laboratory technicians.   (5) Chemical and processing industries pride themselves nowadays on their almost fully automated production lines.   Humans still present have mostly monitoring, maintenance, and supervisory roles, rather than active involvement with the on-line production process.   (6) The introduction of electronic typesetting, including laser-based techniques for font and format changes, has so reduced the bargaining power of the once-powerful printer's unions that they have to rely on seniority-based protective measures to retain a degree of temporary control over their jobs.   (7) Among the most sophisticated manufacturing industries, such as automobiles and aircraft, numerically controlled machine tools are now part of the standard procedures; increasing the degree of computer-aided design and manufacturing is apt to lead to fully automated production plants.[63]   (8) Even that most cherished bastion of the skilled blue-collar workers, the tool and die making specialty, is threatened by the introduction of machinery capable of translating the configuration of the machined product (as defined by other computer-assisted design) into numerical instructions for the preparation of tools, dies, and jigs to be used by automated machinery.

There is no question about these advancements being useful in reducing the direct cost per unit output of production. In fact, the number of individuals in a certain economic sector may even increase in absolute terms, if, as a consequence of cost decrease or improvement in service, the product becomes more widespread than theretofore. But the relative distribution and the nature of skills to be employed will continue to change at a pace perhaps alarming to an increasing number of people. It is perhaps not untimely to start reflecting now about the relatively new social issues involved.

The new elements in this unfolding situation are that it occurs at an increasingly rapid pace and that it affects groups of people with no traditional familiarity with occupational obsolescence. The rate of technological advances is not likely to slow down; in point of fact, owing to the introduction of cheap, reliable and almost intelligent electronics, it is quite likely to accelerate even further. The individuals threatened in their job security are those who heretofore strongly rejected the idea of relying on government help to make a decent living. Feelings of insecurity and the concomitant loss of self-respect lead to resentment and often in a tendency to blame the employer, "big business" and (naturally) the Government. The segments of society thus affected are easy prey to demagoguery; having then their expectations repeatedly disappointed, they become alienated from their peers, friends, family, and society as a whole. Their participation in public affairs, if present at all, remains focused on their immediate interests; their concern for the permanence and cohesion of society, in the absence of clearly present collective threat or impending catastrophe, could be fairly described as minimal.

– NOTE E –

GROWTH OF THE DEBT BURDEN

The Federal Government and, to some extent, state and local governments are often accused of profligacy (unwisely spending the taxpayers' money) and of placing the debt burden on future generations.   This note offers data to support the arguments that (1) Federal budget estimates are notoriously unreliable, being subject to impacts by unforeseen and accidental perturbations (Table E-1); (2) The overall debt level in relation to the "size of the economy" has been essentially stable over the three post-World War II decades (Figure E-2); and (3) The consumer and household debts and corresponding debt service charges by far exceed in absolute value and relative growth those of the Federal and state governments (Table E-3).

The reliability of past and current Federal budget estimates can be assessed from the following table.[64]

TABLE E-1

Fiscal Year	Budget Forecast (1977) (B$)	Actual or Latest Estimate (May 1980)	Increase B$	Increase %
78	440.0	450.8	10.8	2.45
79	466.0	493.7	27.7	5.94
80	496.6	568.9	72.3	14.56
81	527.0	611.5	84.5	16.03
82	558.0	693.3	124.6	22.22

Given this magnitude of errors in budget estimates, the Federal debt management must be described as remarkably cautious.

Figure E-2 below is replotted following Benjamin M. Friedman.[65] Quoting from his findings:

> "Since World War II, the chart shows, the outstanding volume of all borrowing— governmental plus private— has changed little in terms of economy's overall size. In fact, the ratio shows a slight decline. In 1946, overall debt came to 155.9% of gross national product, the broadest economic gauge. The 1979 estimate was 145.5%."

> "The chart further shows that the federal government accounts for a dwindling share of all debt. In 1946, federal debt came to 103.5% of GNP, while last year's rate was only 27.1%. Meanwhile, the consumer's debt

Postwar Borrowing

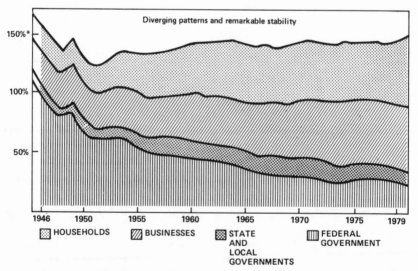

Sources: Federal Reserve Board and Commerce Department
* Debt outstanding as a percentage of GNP

Figure E-2

role has grown explosively. In 1946, such household debt—mainly home mortgages and installment loans—amounted to 19% of GNP. Last year's level, a record, was 53.9%. Debt outstanding among private businesses, at 29.4% of GNP in 1946, reached 52.2% last year."

Note that the "household" category includes both home mortgages and consumer (installment) debts.

Finally a table compiled from the U.S. Bureau of Budget statistics[11] gives an idea of the debt burden growth over the past 34 years:

TABLE E-3

	1946	1980
■ GNP ($ Billion)	209.8	2,627.6
■ Federal Debt (including off-budget items)		
• ($ Billion)	227.0	972.4
• % GNP	109.0	34.0
• Service charge (90-day Treasury Notes) --%	0.375	11.5
• Debt Service ($ Billion)	0.850	111.7
• Debt Service (% of GNP)	0.41	4.2
■ Consumer Debt ($ Billion)	8.78	394
• % GNP	4.2	15
• Interest (avg) %	0.81	15.27
• Debt Service ($ Billion)	0.071	59.9
• Debt Service (% GNP)	0.034	2.28

[11]   Verbal communication from Mr. Vincent M. Helman

Again, while the Government's debt service burden increased 10.24-fold, that shouldered by the consumers (exclusive of mortgages) has grown 67.5-fold between 1946 and 1980.

— NOTE F —

SOCIETAL MANIPULATION

Without the presence of conspicuous external challenges, the military posture of democracies tends to deteriorate. Even though the military and the intelligence community issue warnings about the growing threat, the competition for funding by other desirable objectives and the congenital mistrust of the military, characteristic of modern democracies, act in concert to constrain, inasmuch as possible, the defense-related expenditures. As senior military officers slowly rise to the top, their first-hand experience of warfare becomes more remote and obsolescent, but their views nonetheless carry decisive weight in shaping the structure of the armed forces. When, in response to unexpected challenges, the military fail under such circumstances to deliver a clear-cut victory, the blame is immediately assigned to them--and the distrust of, and disaffection with, "military solutions" become even more widespread.

The key element here is the public perception of the external threat or challenge. It is in relation to these that important changes have taken place since the early 1960's. The invasion, and the subsequent incorporation into the Soviet Empire, of the East European satellites (1944-1948); the Greek civil war (1947); the publicity surrounding the Soviet development of atomic (1949), and later of thermo-nuclear, weapons (1954); and the Soviet-endorsed invasion of South Korea (1950) could be construed as major strategic mistakes by the Soviet Union. They have resulted in the tripling of the U.S. defense budget from 1950 to 1952; they have also cemented the (then quite solid) anti-communist alliances under the aegis of the United States.   The 1962 Cuban missile crisis was the last one on record where the Soviets have taken a significant military initiative, hostile to the U.S. and clearly attributable to them.[12] Ever since, the U.S.S.R., having thoroughly assimilated the lesson of these earlier mistakes, now proceeds with extreme caution-- not a single instance can be found where Soviet "aggression" against Western interests can be clearly and unambiguously proven.   In Vietnam, in Angola, in Ethiopia, and in the Congo, they have encouraged and supported the side favorable to their ideas in civil wars; the Cuban "mercenaries" have acted as volunteers, or at least through the will of Fidel Castro.   In Angola the signals were particularly confusing (1975-1976); it just so happened that the political machinery of the U.S., already benumbed by Vietnam, was further confused by the white/black dichotomy[13], by the hoary old issue of colonialism, the importance of mineral resources and shipping routes (the Suez Canal was still not open)--and as always, by the impending Presidential campaign.

The Soviet behavior is far from perfect from the "manipulative" viewpoint.   They too have pressure groups (the military, the various "factions" in the Politburo); they too are to some extent unable to fully control the conse- quences of their initiatives.   But, on the whole, they have (1) established naval presence in the Mediterranean and in the Caribbean without provoking serious protest or reaction in the West; (2) retained Vietnam and Cuba as effective and willing proxies; (3) preserved their role as major non- Western military equipment suppliers, in particular among the radical nations and groups in the Middle East--and all

12 The brutal suppression of dissidence in Hungary (1956) and in Czechoslovakia (1968) must be viewed as a defensive reaction by the Soviets. (Note added in proof) Even in the presence of such blatant provo- cation against communist orthodoxy as the one now taking place in Poland (August 1980-April 1981), the desire of the Soviets to avoid *open* intervention is clearly apparent.

13 Who would have dared to make common cause with the faction supported by South Africa, the object of hatred to a sizeable portion of the U.S. electorate?

this once again without a single conspicuous action which would awaken the West.

There are knowledgeable observers of the East-West relationships who espouse the view that the Strategic Arms Limitation Treaties (SALT) and the long-drawn-out painstaking negotiations leading to their conclusions are part of a Soviet manipulation scheme. Its purposes, according to the proponents of this view are (1) to shift the actual and perceived strategic nuclear balance in favor of the Soviets; (2) to create uncertainty, and even a feeling of inappropriateness about further massive investments by the West in nuclear weaponry; (3) to preclude Western Europe (and eventually Japan) from possessing theater nuclear weapons in the operational sense; and (4) by driving home the local and global nuclear supremacy of the Soviets, cause the U.S.-centered alliances to distintegrate. If true (and if successful), this manipulative maneuver will have served the Soviet interest very well indeed.

It would be a pity not to comment on the remarkable record of deliberate manipulation of the U.S. Government and population by the expert psychologists of the Islamic Republic of Iran[66] during the hostage crisis (November 1979 through January 20, 1981, 12:35 EST): (1) Release of women and of most black hostages; (2) Extension of Christmas mail and message privileges; (3) Rumors of impending release of the hostages a few days before the U.S. elections and then abruptly quenching the same rumors—sequence which just about clinched President's Carter's defeat; (4) The exquisite precision of the actual timing of the release, which avoided dealing with President Reagan, obtained the maximum concession from the Carter team while denying President Carter the "moral" triumph of having secured the release.[14]

---

[14] Note (paragraph) added in proof

– NOTE G –

FORECASTING IN ANCIENT AND MODERN TIMES

The differences between the tools available to the fore-casters of classical Rome and to those of modern times should not obscure the essential similarity of the purposes of forecasting and of the professional prerequisites for success. The *haruspex*, assigned to a Roman Legion, was expected to advise the *imperator* (with statistically meaningful accuracy) about the outcome of major operational decisions, such as to engage (or to avoid) battle at a certain time or in a certain place; to conclude (or to refuse) alliance with some ill-accoutred, but potentially useful barbarians. His ostensible tools were pitifully inadequate; the entrails of freshly slaughtered sheep, while no doubt revealing inter-esting details in their intricate and unpredictable patterns, have relatively weak coupling to the objective factors which decide, say, the outcome of a given military engagement. More pertinent inputs could be listed, such as the weapons and logistics available to the combatants; their physical shape, their temper, and discipline; the background, competency, the resolve of the opposing commanders; and the availability of reliable intelligence. One comes to suspect that the *real* tools used by the competent segment of professional forecasters must have been quite different. One may surmise that a top-drawer haruspex enjoyed good and confidential relations with his Commander-in-Chief and with the senior officers; he certainly had exposure to their past behavior and the opportunity to assess the strength and morale of the friendly side. He may even have been privy to intelligence sources *not* available to the commanders. Most importantly, he may have subtly detected, when not explicitly ascertained, the mood and the intentions of real decision makers.

Modern forecasting tools, those ostensibly employed and those which are in fact used and are conducive to success, have remarkable similarities to those just described. Few people would go so far as to suggest that sophisticated computer simulation models are no better than sheep entrails in regard to relevancy to the real problems at hand, but contemporary discussions on environmental monitoring and forecasting (in both government and corporate contexts) abound in references to the importance of serving the "climate" of the sponsoring agency, the role of "imponderables," the "commitment to goal-setting," and the "leadership" displayed at the top. To put it more bluntly, just as the haruspex, the modern forecaster may, in many instances, color his predictions by what the sponsors or chief executives wish to hear; or perhaps more innocently, offer a ready rationale, supported by complex, albeit irrelevant, computer models for the decision makers to proceed at what they wanted to do in the first place. The subjective beliefs of the forecasters in the intrinsic value of their tools, sheep entrails or computer models and all those which fall somewhere in between, become largely irrelevant.

— NOTE H —

BREAKING THE ECONOMIC STRANGLEHOLD OF ISLAM

IN THE 15th CENTURY

(Outline for a Case Study)

This note purposes to show that the current economic confrontation between the Moslem and Christian worlds is not the first one; that the particular episode that took place between roughly 1415 and 1550 A.D. had all the potentially lethal geopolitical consequences now being foretold for our society by the most pessimistic prognosticators; and that at least in this instance, the West managed to emerge victorious by a combination of leadership, long-range strategic planning, and (the indispensable ingredient) fortuitous concatenation of favorable circumstances. In less erudite terms the same combination is usually referred to as guts, drive, and just plain dumb luck.

In the first decade of the fifteen century, the Moslem threat was looming large in the eyes of the West. The Middle East,

North Africa, and the Iberian peninsula were firmly in Moslem hands; following the final defeat of the Crusades (1292), the trade routes to the Orient were fully under Arab control.[15]

This trade had at the time the same relative economic importance as oil has in our days; powerful interests were dependent on imported spices, aromatics (myrrh and frankincense), drugs, textiles, and manufactured products from as far as China. The value of these oriental products may have been subjective, but in the aggregate they were sufficient to fuel the major part of the medieval global trade for several centuries.

Control of trade routes in those days was fully equivalent to a monopoly. There was no other known way to reach the Orient from Western Europe, at least not within the cost and the dependability required for stable commercial relationships. The Arab owners of the oriental trade monopoly charged, of course, literally whatever the traffic would bear, with results that have a surprisingly modern ring: (1) Search for technological alternatives (not very successful, in the then primitive state of technology); (2) Balance-of-payment problems, with the attendant scarcity of gold and silver specie in Europe (this was still a world where money had intrinsic value and the idea of reinvestment of surplus capital in the West never would have occurred to any Arab ruler, even had it been technically possible); (3) The rise of "multinationals" (the Venetians, in this case), who were willing, at least in practice, to give up their ideological convictions and sometimes their political allegiances in order to profit from the Mediterranean shipping monopoly, the European financing structure, the distribution network, and even retail outlets.[16]

---

[15] Land caravan routes from the Mediterranean to the Red Sea and to the Persian Gulf, with connecting coastal shipping traffic to India and later to (what are today) Malaysia and Indonesia.

[16] The Venetians played at least an ambiguous role in the siege of Constantinople in 1204, almost 250 years before it actually fell to the Turkish Moslems.

By 1550, the situation essentially changed. The Moslems were still in the heart of Europe; as a matter of fact, they were laying siege to Vienna in 1683, but their "geopolitical" position with respect to world trade had been utterly destroyed. The Portuguese discovered an alternate maritime link to the Orient around the Cape of Good Hope (1498); at the same time they developed the naval power required to establish and to support far distant overseas establishments (factorias, later colonies). The Spanish benefited from the Portuguese advances; based on a cosmogony right in its essentials but utterly wrong in its quantitative conclusions, their Westward drive led them all the way to the Philippines, with the completely unanticipated side benefit of discovering and exploiting the then most attractive prizes of the Western Hemisphere.

It may be presumptuous for any man to claim credit for bringing forth all this revolution; Prince Henry of Portugal ("The Navigator") would have been the last to make such a claim. The facts, as now known, indicate that he personally played an essential role of initiator and planner; his successors, when not preoccupied by dynastic diversions, followed through with vigor and perserverance for more than a century after his death.

After the then compulsory military career, Prince Henry retired in 1420 to his castle at Sagres, overlooking the Atlantic Ocean. His writings and instructions have all the characteristics of a modern strategic plan; he also had the untold advantage of not having to deal with parliaments, Congressional hearings, and other new-fangled noisome devices. His formally professed goal was to serve God by spreading Christianity among the infidels; breaking their trade monopoly appeared as an appropriate means for reaching this goal. Should any wealth accrue to his House by pursuing this course, so much the better. Fortunately for Christianity, he did not have to present long-term cost-benefit calculations ("net present value of expectations") vs. the discounted cash flow requirements. But he did invest in new technologies; in particular, he funded basic sciences, such as mathematics, astronomy, and long-range navigation (at a time when the shape and size of the Earth were mostly unknown, and when indeed the mere fact of speculating on the subject was to invite, fifty years later, the fearsome frown of Torquemada, the Inquisitor). The Prince established schools of naval architecture, supported by competent shipbuilders and well-appointed dockyards. Experimental hull shapes and sailing rigs were systematically developed and tested for speed, handling, and seakeeping. Maritime charts, hydrographic data, sailing instructions were systematically recorded; somewhat later, when the cumulative progress started to pay off in political and economic terms, an almost incredibly efficient security system was established by Henry's successors and maintained for over 150 years in order to deny information of value to possible competitors. What to modern eyes appears most amazing is the detailed competency of the Prince in matters only remotely connected with his immediate purpose[17] and also his realistic perspective on the rate of progress to be expected.[18] He was also very practical and had a good understanding of what motivated his associates. Capital was provided by would-be traders in a legal arrangement that had all the attributes a of joint-stock company, with His Royal Highness being one of the stockholders;[19] his sea-captains were given their posts only after the most exacting training and examinations. He furthermore arranged "incentive payments" and profit sharing that could make a competent and courageous captain wealthy for life in one single lucky sea voyage. Just as important, the Prince was known to be a good Christian and a mild man, but implacable and even cruel when faced with dishonesty or incompetency.[67]

---

17 One well-authenticated effort was a survey of timberland within the realm to supply keels, planking, and staves for shipboard storage tuns; quite remarkably, he ordered reforestation of all the areas clearcut to supply his future shipbuilding program.

18 In his initial instructions, as well as in his last will addressed to his successors, the Prince clearly stated that he does not expect to live to see the fruits of his work (in fact he died in 1460), but that his nation will benefit "to the third and fourth generation and beyond."

19 In this he anticipated Ferdinand and Isabel of Spain by 25 years, but Queen Elizabeth of England by a full century. In fact, since he did not enjoy royal favor or financial support, most of the initial capital was provided from his (rather comfortable) personal wealth.

Soon the results became apparent. By 1540, the Azores were rediscovered; under the rule of Affonso V (1460), nephew of Prince Henry, and eventually under John II (1481), the myth of the Equator's impassability was dispelled soon thereafter (1474); the average southward progress toward the Cape was about 12º latitude per decade. Military and trading outposts were established over the whole West coast of Africa. Valuable information was obtained about winds and ocean currents;[20] a whole generation of sailors, marines, traders, planters, merchant-adventurers set a completely new tone to the Portuguese nation. By 1509, under the Duke of Albuquerque, Goa on the East coast of India and Calicut on the Malabar coast were explored and subjugated; profitable trade relationships were established by 1511 with the Spice Islands (Amboina, Ternate, and Tidore), east of Celebes, Canton (1518), and eventually Macao (1540) on the South China coast.[68]

The direct impact on Portugal was little short of spectacular. In the 16th century, it became the powerful and respected ally of England; its African possessions outlasted the British Empire. Spain, Holland, Belgium, and mostly England benefited from the newly found means to assert maritime power; up to 1970, this characteristic Western attribute was not challenged from the outside.

For the Mediterranean, the results were just as important, although mostly negative and somewhat slower to impact on popular attitudes. In short, from its former status as the political and economic center of the Western world, its significance steadily decreased over the following centuries; by the 19th century, it was little more than a transit route toward the British Empire's overseas possessions. Malta, Alexandria, Cyprus, Port Said, Suez, and Aden were not relinquished until recently, and Gibraltar is still in British hands.

[20] Pedro Alvares Cabral, under Vasco da Gama's command, taking advantage of favorable winds and currents, reached as far as Brazil (1500); by this felicitous discovery Portugal was allocated that country by virtue of the Treaty of Tordesillas (1494) and retained possession for more than three centuries thereafter. The national language of Brazil is still Portuguese.

The impact on the Arab world was downright catastrophic. Threatened from the north by their Ottoman Turk Moslem "brothers" (in fact barbarian and implacable enemies), deprived from their major trade monopolies by the Portguese, later by the Dutch and the British, their civilization was on a path of decay until, in the second half of the 20th century, the oil monopoly allowed them again to reenter the scene of world history as one of the major protagonists.

Seen from the Western European vantage point, the 1410-1550 period was one of aggressive, growth-oriented creativity. The results actually achieved, and even more those rendered potentially possible, were tangible and highly positive. Access to the spice trade, circumnavigation of the globe, discovery of the Western Hemisphere, impetus for exploration and the means for establishment of future colonies were the most obvious manifestations.

It is possible that the same evolution would have taken place, perhaps at a less explosive pace, had the Arab economic pressure not triggered the Portuguese initiatives. But it is intriguing to view the Arab monopoly situation as having actually served as the driving force and the catalyzer necessary to bring about one of the highest growth rates experienced by any continent-sized group in history. The technology potential (navigation, shipbuilding) was present, a generation or two of education and development of industrial base was required (but, with the benefit of hindsight, quite manageable); the essential novelty was the drive and the inspiration of very small numbers of visionary leaders, supported by competent and dedicated advisers, who managed to combine their ambition and wisdom with the executive authority to implement the necessary long-range steps—with no thought of immediate benefits to themselves. The intellectual commitment to a long-sustained effort, with all the necessary components properly integrated, time-phased, and managed, was the mark of true leadership at its best.

Is it possible to view the current OPEC confrontation with the West in the perspective of an analogy? Is it possible that the current pressures, unpleasant as they may be in terms of trade deficits, inflation rates, economic stagnation and unemployment, supply in fact the driving force and the catalyzer for the West to pull itself together and to apply its considerable resources to solve its large-scale structural problems *before* they erupt in cataclysmic consequences? A few areas of endeavor, suggested in Chapters X and XII, undoubtedly offer possibilities; investments aimed at energy independence and reconstitution of trade gradients are likely candidates. But the real point of this note is not to entertain the reader by recounting the anecdotical details of long past history, but to apply its possible lessons to the future. Who, and where, are the inspired leaders capable of thinking through the necessary plans of action? How can they (assuming that they exist) mobilize public and private enthusiasm and substance toward long-sustained efforts? How are they to maintain momentum in the face of temporary setbacks and disappointments? Above all, is there anybody alive who would engage in this type of undertaking not for his immediate satisfaction, but for the benefit of generations as of yet unborn?

_____

– NOTE J –

QUALITATIVE CONTROLS ON INFORMATION FLOW

There are numerous instances in which societies, clearly democratic by Western standards, have undertaken to control or modify the information directed at or available to its citizens. Such public interference, so contrary to the principles of "freedom of speech" and "freedom of press"[21] enjoying Constitutional protection, is presumably justified on the basis of protecting the *recipient* rather than the *originators*. The information or "message" content, format, and reliability are in general recognized as criteria for "quality" assessment.

Some instances warrant clear prohibition sanctioned by law, when the information transmitted is patently dangerous or harmful. The classical dictum of Justice Holmes: "the freedom of speech does not allow shouting 'Fire!' in a crowded theater" is perhaps the most obvious illustration; but other cases, where potential damage to individuals or groups is evident are equally regulated by legislation. Malicious libel and deceptive advertisement are in this category. In other cases, the use of public facilities is denied, e.g., the prohibition of tobacco or liquor advertisement on television.

When potential damage is present, but the case is insufficient or controversial, the Government or private groups attempt to bring some of the facts to the attention of the recipient (presumably prompted by other messages to act unwittingly against his own interest). Thus, gaudy packaging notwithstanding, the supermarket customer may be warned about the content and freshness of the products through labeling and date coding; he may get better value through unit pricing regulations; he may even protect his lungs if he can read the Surgeon General's conspicuously inconspicuous warnings on cigaret packs. The "truth in lending" type

---

21   For technical reasons, the Constitution speaks of "press" only. In recent times, many interpretations have extended the related privileges to all other means of mass communication, including those which use publicly owned media such as radio and television channels.

federal legislation is a remarkable recent instance in which the customer is enabled and encouraged to balance his near-term vs. long-term preferences. He, in general, *wants* the purpose of the loan and *suffers* the burden of high, and perhaps unfair, interest rates, prepayment penalties, late charges, and collection fees. These latter were often hidden in a maze of fine-print details, clever wording, or arithmetic subleties. Needless to add, that the lending profession has vigorously opposed remedial legislation, in particular the unambiguous disclosure of annual percentage rates of interest.

The *quality* of message can be often improved by simply informing the audience about the identity (and by implication, the biases) of the source. This might be obtained through legal compulsion (e.g., sponsorship of political advertisement) or through more-or-less voluntary disclosure by editors, publishers, etc. In the same spirit, the press and public information media can often compensate for biased viewpoints by offering equal time and prominence to both sides of a controversy. In important instances, such as in political contests, the doctrine of "equal time" may be enforced by legislation.

Professional codes of ethics among those responsible for mass dissemination of information are probably the strongest means for controlling the information flow in a democratic society. They are mostly related to moral or ethical value judgments; and cover matters difficult to define, such as good taste, decency, etc. A few years back, it was the professionals who deliberately rejected "subliminal" advertising as invading the psychological privacy of the audience. This may constitute an important precedent.

Finally, public authorities, foundations, and universities (but perhaps to an increasing extent, business organizations) can and should encourage the development and the mass dissemination of "objective" information, including the known facts together with advocacy and adversary statements. [22] The Federal and State Governments publish myriads of specialized pamphlets on topics ranging from alfalfa growing to zooplankton filters; perhaps significant impact can be produced by broadening some topics to address specifically the causes of societal stresses. Here again, the quality of information available to the readership is being deliberately improved.

— NOTE K —

"ACTIVE CONSENT" OF MESSAGE RECIPIENTS

A relatively new and probably controversial concept is discussed in what follows. Briefly stated, it consists of implementing the principle that "freedom of speech" and its extensions must be balanced by the protection of individuals from invasion of their privacy by messages not expressly desired by them.

The basis for advocacy of this principle and of efforts toward implementation rest primarily on the observation that the ability to effectively utilize information ("messages") is contingent upon the message rates being consistent with the recipient's assimilation processes. Since the individuals can, and should, be the sole judges of whether or not they feel comfortable (i.e., able to follow to their satisfaction whatever mental effort the incoming message flow may require), they should also be in position to control the nature of the message flow.

The technical nature of modern information environment (cf. Chapter IV) lends urgency and importance to this aspect of individual freedom. The cost of message generation and transmission has decreased over the past 40 years to the point where the mass of information generated and diffused has become very much larger than that which can be

---

[22] Voters pamphlets published by public authorities in several states follow this format and are, in general, considered most beneficial.

effectively utilized by the population. The available technical means to couple into the individual's brain have become powerful and sophisticated; just about every possible dimension of visual and aural stimulation has been explored and exploited.[23] Violent transient effects, such as percussive sounds and blinding light flashes, are used as attention getters; persistent repetitious and scientifically compounded associations attempt to imprint messages on our subconscious.

If such massive efforts were undertaken for a single political purpose by a dictator, they would immediately be labeled as brainwashing or thought control, with ominous references to Orwell's "1984" (which date, incidentally, is not too far away). If any *one* private interest group would attempt to utilize even a fraction of the total information flow for the purpose of furthering a relatively narrow set of objectives, it would be immediately condemned as engaging in "publicity blitz" or worse; its purpose would, in all probability, be defeated. But just because the originators are many and the motivations are as numerous as they are divergent or contradictory, the process is not only tolerated, but indeed protected.

The several mechanisms just mentioned—distortion, saturation, and manipulation of the information flow—can harm the individuals *directly* in a near-term context by in effect denying to them the basis for rational decisions. More to our central point here, the same mechanisms corrupt the integrity of the political process by precluding the truly informed consent of the citizenry to the actions of the Government. The intended checks by the electorate on the powers of the Government become a concept devoid of true significance, resulting in the stresses and concomitant alienation described in earlier chapters.

If the principle that the right to generate and diffuse information should be balanced by the protection against undesired information is accepted, then the problems of

---

[23] Commercial advertisement is gingerly experimenting with olfactory messages...

practical implementation arise. While public and professional quality control of the information flow [24] shows an increasing measure of acceptance over our history, the principle of "active consent" by the information recipients is relatively new [25] and, on the surface, it might be seen as incompatible with the constitutionally guaranteed freedoms. There might also be legitimate concern about the possible excesses in the direction suggested by the "active consent" principle, i.e., that under the pretense of protection, the individuals, in fact, would be *denied* information they may wish to receive, or should wish to receive; or that the values of some elite group may be used as criteria for acceptance or rejection. These concerns could be mitigated by calling for the evolution of professional ethics among those responsible for the mass communication media, and by allowing in all cases the putative beneficiaries of any "protective" move of the type discussed here to be the ultimate judges as to whether or not such protection is in their interest. In regard to the latter aspect of implementation, there is one important point, though: the consent to receiving information should be *active*, i.e., conscious (although not necessarily costly or burdensome) choice of the recipients. A few examples might illustrate this point:

Electoral Contests: A number of 15-second or 30-second "spots" are used on television to give "exposure" to candidates resulting in (presumably favorable) "name recognition." There is little discussion of substance related to the candidate's qualifications; the statements or promises about future intentions are little more than slogans. The photogenic virtues of all the candidates are carefully explored and, when necessary, artfully enhanced. In most instances, the net result is a large and growing number of competitive and contradictory "spots," where the substance is almost irrelevant, but the images and the persistent repetitions proportional to the available financing and to the

---

24 Appendix - Note J

25 I am not aware of its having been publicly advocated in the form described here. One instance which comes quite close is the right of postal patrons to reject third-class, mass-addressed mail, mostly advertisements and solicitations.

revenue of the broadcasting stations in effect decide between victory and defeat. Much could be accomplished by the simple expedient of grouping all the political "spot announcements" in a single program block during the campaign (called "The Candidates' Hour" for instance) where the short spots of competing candidates would be presented in random succession. It is a safe bet that the futility of this process would soon become evident to broadcasters and sponsors alike, which simply proves the point that the viewers are *captured* and subjected to basically worthless or manipulative information by interspersing the political announcement spots among otherwise desired program segments. The argument that the consolidation of spot announcements would deprive the broadcasters of legitimate revenue could be countered by suggesting that a few serious presentations, say 10 to 30 minutes per candidate, stating their case in matters of both substance and image, including any format (campaign speech, interview, panel discussion, debate), would be just as profitable and probably far more useful to the voters. But here is the essential core of our point: such type of programming would require the *active* consent of the viewers; i.e., switching on, or even *not switching off*, their sets for previously announced political campaign program. The difference is between passive acquiescence (because the short spots are not worth the trouble of switching off the set) and the active consent to receiving the message.

Commercial Advertisements: Much of what has just been discussed is applicable, more particularly in regard to television advertisement. Here at least the profit motive is openly acquiesced; advertisement has essentially lost its etymological meaning ("make aware of") to assume the role of blatant, unabashed attempts to influence. In this place, instead of arguing the content, let us focus our attention on the message format. Shrillness and persistent repetition are the most immediately apparent characteristics; sound effects range from strident to sensuous (dependent on the product) just as the tone of the announcement may be described as insidious, interlocutory or even downright imperative. When ostensibly sponsoring highly attractive programs, such as sports events, concerts, movies, or even political coverage, these are invariably preceded, followed, and interrrupted at frequent intervals by short announcements touting the sponsors' altruism or the product's excellence. Completely unnecessary "station identification" breaks every 15 minutes expose the almost helpless viewers to the relentless onslaught of several commercials discussing their alleged needs and their most intimate concerns. Here again, the large volume of messages reaches the individuals in the audience without their active consent. The sponsors rely on the viewers' mental or physical inertia to leave their eyes and ears open and thus to have their attention (or perhaps their subconscious) captured. If the sponsors and broadcasters are really serious about their desire to serve the public, they could easily put their assertion to test: Let each program event be preceded by a short announcement merely identifying the sponsor. Then, following the uninterrupted enjoyment of the program(s), broadcast in a separate program section (perhaps called the "Consumers' Hour") short features, typically 5 to 10 minutes in duration, explaining in some detail the merits of the products, the virtues of the manufacturers, or the excellence of the service. The ratings attributed to this Consumer Hour type broadcast would soon convince the sponsors of the public's preference. The advertising agencies and their art department supporters would probably find the change upsetting at first, but probably most rewarding in the long term.[26]

News: Here we really discuss both the format and the content. The spread of news wire services has made it economically possible to have immense volume of daily and hourly updated messages, covering the remotest countries in the world, available to the smallest newspaper, radio, or television station. The public has shown an insatiable and increasing appetite for news; to bring the public "up-to-the-minute" news has become the hallmark of service and a decisive factor in competition.

---

[26] Some professional journals of high caliber have adopted this format (grouping all advertisements in separate sections) with no apparent ill effects from the business standpoint.

Owing to the coverage and speed of news services, most outlets can only use a very small arbitrarily selected portion of the news. The selection process is not always wise; snippets are published for their sensation value, often aired without due confirmation, and contradicted later. News commentators do their best to "comment," i.e., attempt to put in perspective, but they too are constrained by the pace and the theatrics of the news shows. It follows that the public may be "informed" in the strict sense of the word, but many messages are irrelevant to most recipients ("mild earthquake leaves 200 homeless in Peruvian village"), some others are distorted, truncated, or simply untrue. Here we come to the format: for reasons of economy and in an attempt to reach as many people as possible newspapers, but even more, radio and television stations, publish and broadcast the same information over and over again in the course of the day until it acquires the authority of fact through sheer repetition. Perhaps the public would be better served by carefully weighing, analyzing, and consolidating the news into meaningful stories, not necessarily up to the minute, but also mercifully emended from all known distortions and unnecessary sensationalism. As in many other instances, it is quite possible that news media taking this approach may well be rewarded by both public approval and profitability.

– NOTE L –

TECHNICAL PROGRESS AND SOCIAL ATTITUDES

Sanitation, health care, and correctional institutions offer instructive examples. As long as the immediate causes of epidemics and infectious diseases were not understood, sanitation was more or less a luxury; health care was, at best, intended to alleviate charitably the most painful symptoms but, in general, of very little real help in comparison to the spiritual relief offered by the ministers of religion. As to correction, even for trifling offenses, the approach was often close to the "final solution" (later endorsed by the Nazi regime for social misfits); maiming, branding, transportation, and hanging were broadly accepted or even approved—rehabilitation was not considered as desirable or cost-effective. Pain and suffering being part of the unavoidable human destiny, its presence was not considered unusual or unjust, especially when brought about by clearly asocial behavior.

The contrast with the late twentieth century is, to say the least, striking. The progress in medical sciences, (Jenner, Pasteur, Semmelweiss, Koch) offered a rationale for large-scale sanitation projects, the building of hospitals, sanatoria, medical schools, as well as the professionalization of the related specialists. Soon, public attitudes toward health care changed rather radically; first, even ordinary citizens expect their everyday surroundings to be wholesome and sanitary, to an extent inconceivable even among the pampered classes of the 18th century and before. If, in spite of all these preventatives, they happen to be sick, not only are they entitled to competent medical and supportive care (with the cost impact largely mitigated by public and private insurance plans), but they definitely expect to get well; avoidance of pain and suffering is essentially taken for granted.

But the most fascinating aspect of this evolution is its impact on social attitudes in regard to correctional institutions. As disease and untimely death become more a matter of explainable, preventable, and remediable incidents, as the cruel scenes of suffering at home and in crudely equipped hospices recede from public consciousness, the barbarous treatment of criminals is no longer accepted as morally right or socially effective. Torture, maiming, flogging have practically disappeared from all advanced nations. Capital punishment, to the extent still tolerated, is surrounded by the most elaborate legal safeguards in favor of the convicted criminal, and at any rate, it has definitely lost its former standing as an accepted mechanism of social interaction. At the other end of the spectrum of asocial behavior, not less than the authority of the U.S. Supreme Court is required to define modalities of corporal punishment in public schools (size and weight of the paddle, thickness of garment veiling the impact area, presence of witnesses); in Sweden the law even regulates the correctional actions by parents toward their offspring. This is a far cry indeed from 18th century England where preteenage children were actually sent to the gallows for what today would be at best adjudged as petty larceny.

Social attitudes *do* evolve.

— NOTE M —

THE ROLE OF GOVERNMENTS IN INTERNATIONAL TRADE

When trade among nations involves discretionary surplus commodities or products from the exporters and the satisfaction of nonvital needs of the importers, or again when foreign trade is a relatively modest element in the participants' economy, then such trade is, in general, left in the hands of private individuals or corporations. Political power, in the form of governments or rulers, enters the picture only to collect customs and excise taxes or to regulate trade in view of favoring certain national activities through restrictions, quotas, and monopolies. As a tool for eliminating foreign pressure or blackmail, most totalitarian nations which consider the satisfaction of individual desires of its citizens subordinate to the interests of the state, resort to government-imposed autarky. Napoleon's Empire, Hitler's Third Reich, and Communist China in the early 1950's are notable examples.

Since the 1960's some essential aspects of world trade have undergone significant modifications. The U.S., Western Europe, Japan, and other Western nations have experienced a growing dependence on fuel and raw materials not available at reasonable cost (or not available at all) within their own boundaries; foreign trade among these nations increased to almost 23% of their aggregate GNP by 1974; they were strongly competing with each other for capital-goods export markets; the U.S., Australia, New Zealand, and Canada remained also dependent on export markets for their remarkably productive agriculture—all this in a period of massive instability of exchange rates, trade balances, and (in the Third World) political systems. As a consequence, political philosophy notwithstanding, each Western nation, in keeping with its own style of government, has in effect resorted to government planning, control, and *de facto management* of their respective economies, most emphatically including foreign trade.[27] This is not the place to discuss in detail the specific implementation aspects of this monumental undertaking, but the following conclusions appear justified:

● The growth and the general acceptance of governments' role in the "management" of the economy seem to be independent of national characteristics or the political coloration of the parties in power; differences are a matter of degree, rather than of principle.

[27] The MITI (Ministry of International Trade and Industry), together with the informal but very real government control on the major financial institutions, all but manages Japan's trade and investment policies. In France, the Ministries of Planning, Trade, Development, and the nationalized energy sector in fact manage the nation's economy; their semiprivate Credit National is the Treasury's instrument for channeling capital investment to the economic sectors favored by the Plan. The U.K., West Germany, and other industrialized nations have comparable institutions. The planned economies of some of the "Popular Democracies" of Eastern Europe, patterned after the successive Soviet 5-Year Plans, are known to be authoritative and, on the whole, more successful than their Soviet counterparts.

● Efforts to manage national economies exhibit the recognized pattern of inefficiencies associated with all government activities (not more and not less). In addition, the problem of long-term management of economic systems in a democratic system is fraught with fantastic complexities, with no valid theories and little relevant empirical knowledge. Perhaps an accurate way of phrasing would be that the *major determinants of the problem evolve faster than our degree of understanding.*

● While considerable lip service is paid to international cooperation among the Western nations, theory and experience is equally lacking in that respect as well; furthermore, the domestic political pressures do prevail invariably when pitted against longer term international considerations.

The innumerable international agencies appear to handle relatively specialized segments of the overall problem. In spite of their prestigious title and ambitious charter (a very limited exerpt given in Table I below), their primary function is the gathering and diffusion of statistical information and the conduct of ad hoc negotiations on issues as they arise. Only exceptionally do these and similar organizations take the initiative for bold policy recommendations affecting the senior (Western) participants.

TABLE M-I

INTERNATIONAL ORGANIZATIONS[69]

The United Nations

● Economic Commission for Europe (ECE) - 1947

Representatives of all European countries and of the United States and Canada study the economic and technological problems of the region and recommend courses of action.

- International Bank for Reconstruction and Development (IBRD) (World Bank) - 1945

    Initially it was concerned with post-war reconstruction in Europe; since then its aim has been to assist the economic development of member nations by making loans where private capital is not available on reasonable terms to finance productive investments. Loans are made either direct to government, or to private enterprise with the guarantee of their governments.

- International Development Association (IDA) - 1960

    Affiliated to the World Bank, IDA advances capital to the poorer developing member countries on more flexible terms than those offered by the Bank.

- International Monetary Fund (IMF) - 1945

    The IMF was established to maintain stability in international currency rates. It has various arrangements for the sale of foreign exchange to countries in balance of payments deficit. The Special Drawing Account was introduced in 1970 as a means of strengthening national reserves.

- United Nations Educational, Scientific, and Cultural Organization (UNESCO) - 1945

    UNESCO was established "for the purpose of advancing, through educational, scientific, and cultural relations of the peoples of the world, the objectives of international peace and the common welfare of mankind."

- Conference on Trade and Development (UNCTAD) - 1964

    It aims to evolve a coordinated set of policies, to be adopted by all its member states, designed to accelerate the economic development of the developing countries. UNCTAD's concern covers the entire spectrum of policies in both developed and developing countries which influence the external trade and payments and economic development of developing countries.

- Capital Development Fund - 1966

    It assists developing countries by supplementing existing sources of capital assistance by means of grants and loans; assistance is directed towards the low-income groups in developing countries who have not benefited from earlier development efforts; assistance may be given to any of the member states of the UN system, and is not necessarily limited to specific projects. The Fund is mainly used for the benefit of the 25 least-developed countries.

- Industrial Development Organization (UNIDO) - 1967

    By resolution of the General Assembly, it assists in the industrialization of the developing countries through direct assistance and mobilization of national and international resources.

European Free Trade Association (EFTA) - 1960

EFTA's object is to bring about free trade between member countries in industrial goods and an expansion of trade in agricultural goods.

Organisation for Economic Cooperation and Development (OECD) - 1961

Replaced the Organisation for European Economic Cooperation (OEEC). The aims are to promote economic and social welfare throughout the OECD area by assisting member governments in the formulation of policies designed to this end and by coordinating these policies; and to stimulate and harmonize its members' aid efforts in favor of developing countries.

Organization of American States (OAS) - 1948

- Inter-American Economic and Social Council (IA-ECOSOC) - 1970

    The principal purpose is to promote friendly relations and mutual understanding between the peoples of the Americas through educational, scientific, and cultural cooperation and exchange between member states, in order to raise the cultural level of the peoples.

- Inter-American Nuclear Energy Commission (IANEC) - 1959

    IANEC advises and assists member states in developing and coordinating research and training in nuclear energy. In addition to providing direct aid to Latin American institutions for work in development and research, IANEC also sends professors and researchers, finances the development of courses, and defrays expenses of Fellows in the training centers. It also distributes information and recommends public health measures.

— NOTE N —

STABLE ECONOMIC GROWTH-ORIENTED
INVESTMENT PROJECTS

The purpose of this note is to record, without in-depth exploration or determination of soundness, some of the ideas encountered while investigating this subject. Some of them are old; several have been suggested by more than one source. Originality or paternity is not claimed by anybody to the writer's knowledge. Some may turn out to be utterly unsound, or may not be conducive to achieving the intended purpose.

1. Topographical changes[28].—Development of artificial ports, harbors, lakes, beaches, reefs, or causeways; improvements in transportation; potential for recreation and aquaculture.

2. Improvement in transportation and utilities.—Renewal and strengthening of railroad beds, underpasses, bridges, tunnels, and waterways. Use of tunnels for multiple utility ducts, including optical communication lines.

3. Systematic mineral wealth exploration.—Provide equipment and information processing system to assess availability of mineral resources in areas not accessible to road transportation. Integrate remote and proximal exploration.[70]

4. Reuse of Water and Silt at River Estuaries.—At river estuaries in flat coastal regions it may be economical to install facilities to recover and pump upstream irrigation water. Silt recovered at the same locations may be used to compensate for erosion and increased salt concentrations in arid irrigated areas.

---

[28] All items applicable to both domestic and LDC-oriented investments (Chapters X and XII).

5. <u>Microclimate Control.</u>—Mass produced, continuous, transparent, tunnel-like structures to provide controlled environment for crops. The purpose is water conservation, controlled application of fertilizers and insecticides, increase in growth season's duration.

6. <u>Nuclear and Reusable Energy Sources.</u>—Nuclear, solar, wind, and bio-mass fueled power plants, operating with minimal technological sophistication and supervision.

7. <u>Improved River Navigation.</u>—Locks, lighters, container elevators to circumvent obstacles to river navigation such as cataracts.

8. <u>Low Infrastructure Air Transportation.</u>—Airships to ensure air transportation of people and freight in areas where the capital investment in modern air transportation is not required or not (yet) economically justified.

9. <u>Oxygen-Free Storage and Transportation.</u>—Nitrogen-pressurized facilities and transportation equipment for protection of grain and other foodstuff against rodents.

10. <u>Information Dissemination.</u>—Satellite-based communications, education, weather services.

– NOTE P –

EMERGENCY PREPAREDNESS AND CIVIL DEFENSE

The role of civil defense in the evolution of the U.S. strategic posture vis-a-vis the Soviet Union has received much increased attention in recent years. The thrust, the magnitude and the organization of civil defense efforts are still subject to intense debate, but the relatively urgent necessity for action is accepted by influential elements of the U.S. Government, and increasingly perceived by the public.[29]

Protection of the U.S. civilian population against the effects of a major nuclear attack is the foremost civil defense objective. The practical impossibility of such a task is largely recognized. The orientation of civil defense, in the minds of many people, is therefore mostly aimed at increasing as much as possible the recovery capability of the U.S. population with its industrial and economic activity. The need for recovery (and indeed the functioning of the nation as an organized society during periods of intense perturbations) is taken for granted, even within the context of nuclear wars.

Nuclear attacks on the U.S. population and industry are probably foremost in many people's mind when thinking of civil defense. The importance of lesser levels of disasters, including but not limited to collateral damage following other types of nuclear attacks, should be recognized. One should also keep in mind the large number of other "disaster" scenarios, having no direct relation to nuclear wars, but which in terms of cumulative probability may represent a threat of the same order to the stability and well-being of society. The purview of "civil defense" should thus be broadened to encompass all the plausible perturbations likely to affect this country in the foreseeable future. Just as military preparedness requires resources and organization

---

[29] Several industrialized nations, especially those that highly value their neutrality (Sweden and Switzerland), are way ahead of the U.S. Others appear to take the matter more seriously than does the U.S. to date.

to face many types of well recognized war "scenarios," civil defense should be prepared to deal with many plausible types of disasters or emergencies. Those most often mentioned are the natural disasters, major industrial accidents, riots, and terrorism. One category which should receive special attention is the interplay of industrial accidents with natural disasters, such as those exemplified by nuclear reactor damage caused by earthquakes or, at a lesser level, regional failure of power grids caused by electrical storms. The relationship between "industrial vulnerability" and terrorism or covert sabotage, especially as it may affect nuclear facilities, hydraulic and thermal power plants, and utility distribution networks, should also be seriously considered.

In many instances of this type the damage is caused by the effects of the disaster itself, conjoined to the extreme fragility and complex interdependence of modern industrial societies. We are in effect dealing with a relatively new phenomenon aptly defined as *societal vulnerability*; conversely, what civil defense and emergency preparedness are primarily attempting to improve and accomplish, in cooperation with other efforts in this general direction, should be described as *societal resilience*. As the complexity and the fragility of a modern industrial society increase, the efforts aimed at increasing societal resilience should be recognized and developed in the long term perspective. They should receive serious sustained attention of the Government at all levels, and, most importantly, they should be integrated into the basic long-term structure of national requirements.

The idea that the resources required to satisfy a broadly recognized, long-term societal need should be integrated into the economic structure (or, in the accepted economic terminology, internalized), is far from being new or original. Examples can be readily found in recent history: The cost of industrial safety, the cost of environmental protection, and the cost of "full" employment are by now essentially internalized in the U.S. and the corresponding financial burden is distributed over the national economic base. The same holds for all other industrialized nations. Civil defense, in the broad sense defined here, is a societal characteristic which is increasingly perceived as an imperative necessity. If some of the more extreme scenarios come to pass, it may well prove to be literally vital to national survival and recovery. Even while the need for civil defense, or more generally societal resilience, is accepted, the decisions at the national level face a number of serious difficulties. There is no organized constituency now speaking for civil defense, neither is there an institutional basis to support it within the Government. The responsibilities of federal and local governments are often confused and conflicting. Many people, remembering the 1950's and 1960's, look upon civil defense as still another exercise in futility, or worse, a ploy by the Government to encroach even more on the overburdened population. Civil defense is also a politically sensitive topic; often and irrelevantly mixed with the misleading dichotomy of war-deterrence vs. war-fighting policy objectives. None of this encourages or promotes a dispassionate and purposeful evolution of decisions required to bring forth a truly efficient civil defense program.

One of the powerful means to accelerate the acceptance of a relatively new and obviously burdensome public undertaking is to seek out the synergistic elements in the new proposed activity. If important aspects associated with civil defense can be shown as contributing to the objectives of other recognized societal endeavors, civil defense will appear more desirable and more acceptable than if it were perceived as just another new contender for limited public funds. If civil defense is organized and publicized so that in the public perception the total value of civil defense and other interacting programs is significantly greater. than the sum of the parts, then it is likely to receive the public and Congressional acceptance and support which are preconditions to its ultimate success.

Based on these considerations, the interactions of civil defense (and perhaps broadened to include the concept of "societal resilience") with other recognized and accepted societal objectives should be explored. It must be accepted as *axiomatic at the very inception of this thinking* that we should think in terms of intrinsic, objectively demonstrable synergies. Any lack of genuine substance, any attempt at public relations type manipulation will be unavoidably exposed and resisted.

The following are a few instances that readily come to mind in the perspective of interaction between civil defense and other societal objectives:

a.   Strengthening the National Guard

Many states have strong interest in giving the National Guard a meaningful role, in particular in the context of emergencies. The equipment, training, and organization of the National Guards should be modified and increased to reflect their civil defense responsibilities.

b.   Stockpiling of Strategic Materials

This program could be expanded to cover not only raw materials and fuel, but also semi-finished products and consumer goods that would be required in case of national emergencies. Industrial products, including machine tools and other long-lead items for recovery could also be stockpiled. The expanded stockpile program can and should be used to stabilize as necessary the otherwise undesirable economic fluctuations. It may become a powerful government tool to control inflation and to alleviate unemployment.

c.   Industrial Dispersion and Protection

The Government should encourage industrial dispersion, especially when it can be associated to desirable shifts in land utilization and planning of metropolitan agglomerations. New industries should be encouraged, by fiscal and other measures, to invest in alternate energy and raw material supplies and in emergency protection facilities. Hardened autonomous powerplants, with underground fuel storages, may be particularly desirable.

d. Changes in Building Codes

The Government should stimulate the redrafting of building codes to encourage societal resiliency. The construction of underground buildings for industrial and residential use should be encouraged. This would save structural costs, heating and cooling energy, would help shelter construction or designation. New buildings should have emergency exit/access facilities as well as emergency power and protected air supply.

e. Emergency Equipment Manufacturing

In any civilian emergency, up to and including nuclear attack environment, evacuation and life support of the population become of paramount importance. Equipment apt to support such operations should be manufactured, stockpiled, and *locally distributed*. The essentials are transportation, food, medical and surgical equipment and supplies, sanitation, communication, and utilities. Emergency housing and individual protection are especially required. The Government should call on industry for the design and manufacturing of equipment to support these needs. Specifically, prefabricated housing and shelters, utility trucks (with fuel reserve), transportable road repair materials, bridges and pontoons will be required. Utility vehicles, perhaps patterned on camping vans, including fuel and food supplies, should be stockpiled. Helicopters should be made locally available together with training programs for helicopter pilots. Alternatively, all-terrain utility trucks would play a significant role. We should also think in terms of self-contained water supply, sterilization, decontamination, heating, and sanitation units being capable of efficient use with prefabricated housing. All these are obviously only examples. The point is that, by identifying and encouraging the development of the right kind of equipment, the Government can use the civil defense needs to stabilize, to a large extent, major segments of industrial activity. The economic impact in terms of stabilizing the demand/supply balance, especially in chronically stagnant industries, is readily apparent.

f.   Use of Agricultural Surpluses

Instead of just being stockpiled or dumped on foreign markets with Governmental subsidies, some of the U.S. food surpluses could be absorbed in the production of emergency food rations.   These should be distributed and *locally* stockpiled, together with the organization of an in-place distribution system.   The stockpile should be continually recycled by distributing an appropriate fraction to the population which otherwise would be the beneficiary of food stamps or similar aid programs.

g.   Use of Labor Surplus

Instead of the notorious make-work type programs, a large portion of the U.S. labor surplus, in both the skilled and unskilled categories, could be put to useful work in the civil defense area without competing with the private sector.   The building of shelters, access roads, and other infrastructure facilities in the proposed evacuation areas are prime examples.   The provision of industrial hardening facilities (access roads and tunnels), the building of bypasses for major road and rail interchanges, the building of emergency airfields are typical examples.   The building of tunnels along access or exit roads to urban areas not only saves land surface for more valuable peaceful uses, but also potentially protects the transportation and utility functions, while also providing shelter space for the population and for its essential supplies.

In conclusion, if the requirement of broadened civil defense is accepted, it becomes apparent that the practical and efficient evolution of a societal response requires the tie-in of civil defense activities with other, socially useful, endeavors. Further exploration and implementation of these synergistic interactions is recommended.

— NOTE R —

INTERFACE EQUIPMENT

In general, any production process calls for the interaction between humans and machines to a degree dependent on the sophistication of the task and the maturity of the technology state of the art. Thus, hand tools and primitive machine tools require considerable admixture of human dexterity and training in order to result in acceptable production quality at affordable cost.

In the recent past, and mostly since the advent of reliable micro-electronic technology, extremely complex man/machinery has been introduced in order to increase the productivity of relatively highly skilled people. A typical instance is the electronic word processing field where electrical typewriters, electro-luminescent displays, computers with magnetic memories are used in combination with highly trained (human) typists in order to improve the purely mechanical aspects of the intellectual production process.[30]

The impact on the work force, once complex equipment is introduced, has several components:

- The most highly educated (or adaptable) segment clearly benefits; this is the one which designs and manufactures the new type of equipment. It, furthermore, gains critical leverage in the new production process by laying claim on operational tasks such as setup, adjustment, maintenance, repair, modification, and the ensuring of compatibility with the (everpresent) growth versions.

[30] There is very little doubt that in the majority of routine business-related typewritten or printed material, electronic word processing often introduces order-of-magnitude improvements in quality and quantity. In the field of creative writing, especially for material of high analytical or technical content, the improvement is probably much less. There is a lot to be said for the relatively laborious longhand or direct mechanical typing—the very cumbersome nature of the process forces authors to think carefully before setting their ideas to paper. In the subsequent phases of editing, electronic word processors would, of course, be invaluable.

- Also high on the skill scale are those responsible for the routine operation of the new type of equipment; in the case of electronic word processors, these would be the typists selected and trained to become proficient in this new production process. Since their output rate is greatly enhanced, one would expect a substantial increase in the corresponding pay scales, and in many instances this expectation is, to some extent, fulfilled.

- Again, in general, by the introduction of more efficient and capital intensive production processes, a number of hitherto skilled or semi-skilled occupations become downgraded or eliminated. The introduction of electronic typesetting has already been cited as an example.

The concept of *interface equipment* (perhaps an emotionally more attractive designation should be found) is based on the assumption that, as new technology driven by the state-of-the-art advances and competitive considerations are being introduced at an ever-increasing rate, a growing number of workers find themselves in the third category. Their competency relative to the first two groups is insufficient; their hold on their current jobs is insecure or the market price for the (down-graded) jobs no longer fills their needs or expectations.[31]

---

[31] The problem shows strong analogy with the "Problem of the Commons" described by Hardin.[71] When a common grazing pasture is exactly in equilibrium with the size of the herd, the situation is unstable. The immediate utility for any one individual in the village to add one more head of cattle far exceeds the long-term ecological damage to the pasture due to overgrazing. Since each villager thinks the same, catastrophe is the logical consequence, unless collectively agreed restraints are applied in time.

The same situation exists in the field of technical innovation. The utility of major technical advances in the production process is undisputable for the innovator; they allow him to gain direct, immediate advantage over his competition, while the "ecological" damage, in the form of increased unemployment, general disaffection of the workers directly affected or perhaps of whole class of workers is far less direct or perceptible. So the process goes on, emulated by competitors, unless, again, collectively agreed restraints are applied.

Interface equipment is defined as that which is deliberately designed to accomplish relatively sophisticated productive tasks by interfacing with human skills corresponding to much lower level of sophistication. Its purpose is to increase the number of workers, at relatively modest skill levels, who can participate in advanced production processes.   By properly designed interface equipment, the downgrading or elimination of jobs caused by technical advancements can be presumably reduced or even avoided.   Point-of-sales (POS) type computer terminals coupled with automatic inventory controls in supermarkets offer an excellent illustration.   In point-of-sale type terminals electronic sensors pick up the product code and the unit price so that checkout personnel use practically no mental effort in the tallying of products. The associated computer also automatically checks on the amount of change, prints out a record of the whole trans-action,   and   verifies   the   customer's   identity,   checking account, and credit card.   It also interfaces with inventory control and prints out new supply orders whenever stocks fall below predetermined thresholds.   Positions at checkout counters now become available to personnel with extemely limited training, provided that they satisfy the obvious minimal requirements of punctuality and integrity.   This innovation, although initially proposed for cost reduction and increase in efficiency, in fact is fully satisfactory from the viewpoint of providing jobs for people who otherwise would not have qualified.

Another example could be found in the solid waste disposal area.   Typically, a large volume of unsorted solid waste is delivered to a station.   The problem is how to recycle the valuable materials, sort them out by category (glass, metal, etc.), separate the organic fraction which is appropriate for fuel generation, and compact all the rest into a form acceptable for sanitary landfill.   If the proper sanitary precautions are taken, individuals with relatively low skills and with addition of only low-level technology can substan-tially contribute to the effectiveness of the process at most affordable capital costs.

Manufacturers of equipment involving complex wiring har-nesses are in the process of experimenting with computer-aided control of assemblies. Here, by means of display screens or indicator lights, the computer prescribes the next step in the assembly to the operator, then automatically verifies the conductance and the insulation of the step just performed, and proceeds to prescribe the next step. Individuals with minimal skill can hold jobs of this type which in the past have required several months of assiduous training.

It may be worthwhile to suggest that major technical innovations should be examined, among other considerations, from the viewpoint of impact on the potentially displaced "marginal" workers. Incentives in the form of tax reduction, subsidies, or contract preferences should be given to industries or techniques which ensure minimal disruption of the labor force.[32] Institutes of learning, both at the university and vocational level, should be encouraged to explore the potential and implication of interface equipment as just defined.

---

[32] The current popularity of "robotics" can be seen as still another step in the direction of eliminating repetitive manufacturing and assembly jobs. Thought should be given to the concept of human-robot cooperation where the (relatively cheap and remarkably effective) sensory and cognitive capabilities of the workers are coupled to the strength, precision, environmental tolerance of the robot(s).

— NOTE S —

THE EMERGING "THIRD WORLD"

We now turn our attention away from the traditional pursuits of international politics: balance of power, spheres of influence, trade relationships, local or regional threats or liquidations of crises...let us think in global terms for a moment. The year 2005 is, after all, only as far away in the future as the first Eisenhower term is in the past; and behold, how many predictions of the 1950's were, in fact, worse than inaccurate or false—they simply became almost irrelevant. Most "futurologists," especially those with a tinge of fascination with technical gadgets, invariably focussed their attention on the technological possibilities (conveniently eschewing the economics), but the broad structural problems of the national and global societies were often ignored, in both senses of the word.

By A.D. 2005 the West will comprise 12% of the 5.5 billion world population; the Soviet bloc about 10%; China 25%; India, Brazil, Mexico, and Indonesia together about 35%; with Southeast Asia another 5%, Black Africa 5%, and others the balance, or about 8%. The exact numbers are unimportant, but the ineluctable fact stares us in the face: About 75% of the world's populations will still be non-Western, nonindustrialized, and probably nondemocratic, even if the Soviet Union refrains from exacerbating the potential crisis situations. The West and the Soviet bloc together, with 22% of the world population, will control almost 80% of the world industrial production and consume 82% of all industrially produced energy. A full 47% of their aggregate foreign trade will be with nations outside of the industrialized group. The West and the Soviets are thus the primary producers and consumers of the early 21st century; the other 75% of the world remain the "hewers of wood and the drawers of water,"[33] or in more modern terms, the

---

[33] Note that even the Old Testament's words use energy and resoure related images to describe the essentials of political subjugation.

"villages" which provide for the needs of the "cities."[34] This North-South cleavage is the essential characteristic of our near-term future; compared to this, the recent and current preoccupation with the East-West antagonism may become an almost insignificant sideshow of history. The latter has acquired importance only because of the self-inflicted weakening of the West, the ideological commitment and military power of the East, and the specter (all too real and seriously threatening to materialize) of nuclear devastation.

The teeming masses the Third and Fourth World[35] will continue what they have been doing for the past couple of decades; they will argue, fight, starve, and (occasionally) vote, but mostly seek food and shelter, and will enthusiastically reproduce whenever the slightest biological opportunity arises...in short, they will continue their "teeming." Even though by some fortunate quirks of geology, a few of the LDC's may accumulate almost obscene levels of material wealth (and in the process endanger not only their souls but also the stability of their governments), their per capita production and consumption levels will decrease on the average. The present trends show that, efforts toward industrialization notwithstanding, the population growth in the nonindustrial societies will outpace the increasing availability of food, jobs, and consumer goods, production levels, or any other measure of material well-being.

As long as the lowest population strata in the LDC's do not sense the true basis of social security (i.e., provisions for old age secure against inflation and political upheavals), they

---

34   The late Chairman Mao-Tse-Tung (old-style spelling), was among the first to describe the North-South polarization in terms of "cities" (the soft oppressors) supported by the "villages" (virtuous, industrious and, possibly therefore, oppressed).

35   The original term "Third World" was coined c. 1948 to designate the group of nations which has chosen to remain "unaligned." "Fourth World" is a relatively new concept; it designates nations without readily exploitable resources beyond those strictly necessary to maintain their population at the barest subsistence level. To avoid any permanent political label, the term LDC (less developed country) will be used in this section to designate both "Third" and "Fourth" World nations.

will obey their fundamental urge to produce offspring and to try (under almost desperate conditions of abject poverty) to raise them to maturity.   India and Latin America offer convincing evidence that compulsion and propaganda will not overcome the population growth; only a couple of generations of political and social stability, with the concomitant growth in per capita disposable wealth, will produce an actual significant slowdown.   This means that the LDC's will not only include 75% of the world's population, they will comprise a frightening 85% of the world's population between 18 and 25 years of age, i.e., the prime reproducing and fighting group.   Any ideology, any leadership, any risk in order to gratify the irrepressible desires fundamentally stoked by biological and physiological pressures will be eagerly accepted.   The superimposed cultural veneer is just that: a superficial coloration to help and to intensify concerted group action but not the fundamental driving force.   This basic fact must dominate the Western policy formulation at the present juncture.

The LDC's are not, by any definition, a homogeneous group. Racial, tribal, cultural differences are probably more widespread than among the industrialized nations, [36] and they are sharpened here by the presence or the vivid recent

---

[36] This assertion may have to be qualified. Scandinavian nations and Japan offer examples of racially homogeneous populations under relatively prosperous conditions; there are, of course, competitions between social groups, but these are attenuated and sublimated under the guises of negotiation, consensus, and "working coalitions." There are some indications that the relatively stress-free societies of this type hide or prefigure a new type of benevolent dictatorship, based on lifelong conditioning and nurturing toward social conformism, associated with the state-sanctioned satiation of most individual, not to say intimate, desires.[72] On the other hand, there are many instances where racial, ethnic, or regional and religious differences persist and are exacerbated by other causes of social stresses such as (relative) underdevelopment, poverty, and discrimination. Conflicts are often present and are smoldering under the careful eye of the state police when not erupting in overt rebellion and civil war. Let those who smugly contemplate the travails about racial and ethnic minorities within the U.S. turn their attention to the Basques, the Bretons, the Irish Catholics, the Walloons, the Flemish, and innumerable others, all only too ready to take to arms to correct injustices of several centuries of standing. An African historian, a half century hence, will write, no doubt, of *white* tribal wars in this context.

memories of humiliations and physical sufferings. For whatever cause, there are enormous social and material gulfs between the sleek Kuwaiti citizen[37] and the starving peasant of Bangladesh trying to literally scratch a living out of a flood-devastated, parasite-intested, ungrateful land to which he is entitled only because nobody else within convenient migrating range is starved enough to dispute his claim.

Even within any one of the LDC's, social stratification is apt to be sharper and more cruel than that in industrialized countries. The old adage "man is a wolf to man" is nowhere as true as in the newly emerging nations. Everywhere, from the Middle East, to Latin America, to Black Africa, to Southeast Asia, the wealthy ruling classes use all the traditional levers of subjugation—faith, custom, ignorance, superstition, indoctrination, starvation, stunting of juvenile development, or even brutal repression—nothing is too immoral or too painful, provided only that it be effective... as it invariably is. In addition, modern techniques add their power to the process: The elites use their accumulated wealth to further their differential survival through overt or covert polygamy; their material well-being through the modern paraphernalia of airconditioned palaces, winter sports, and the miracles of Western medical care; their security by means of praetorian guards equipped with the most modern weaponry and security systems. Just in case all these precautions would prove insufficient, numbered bank accounts in Switzerland with comfortable credit balances are considered *de rigueur*. Then, and then only, when all these essential needs are amply satisfied, will the ruler condescend to cater to relatively modest (although lavish by past standards) programs of social security, health care, and a very modest program of "universal" education. The female half of the population is still used, for the very

---

[37] Or, for that matter, the lucky denizens of Abu Dhabi, Qatar, or other powerful nations, which would still have a GNP growth fueled by camel dung, except for the Western invention of, and idolatrous fascination with, the internal combustion engine.

same reason, as submissive breeding stock, since it is well known that "a woman always knows too much for her own good;"[38] besides, once a woman is given knowledge power, she will invariably and effectively use it for subversive purposes.

Western policies must be definitely based on premises which recognize not only the widespread lateral diversity of the LDC's, but also the cultural and economic stratification which is in the process of emerging *among* the individual nations, and the cultural, social, and economic divergencies which are in the process of growing harsher and more visible among *the various segments* of the individual nations. While the rhetorics of foreign policies (mostly originating among the elites speaking for the LDC's; the others do not have the travel funds to participate in international symposia) will argue the cultural values, the global redistribution of wealth, and other such lofty principles, the *community of interests among the less favored strata* of the LDC's is growing; it is likely to transcend cultural, racial, and national barriers.   The "grapes of wrath" are rapidly maturing;[39] thanks to the rapid diffusion of information,[40] the maturation process is likely to be much faster and much more violent than in the past evolutionary phases of history.[41]

---

[38]   The direct quote is from Moliere (1672), but it is a safe surmise that others (mostly men) have expressed the same thought before, and ever since.

[39]   Steinbeck.

[40]   Cf. Chapter IV.

[41]   There is ample evidence to support the relationship between the information flow (as it reaches effectively the lower population segments) and the rate of societal evolution. (1) The stability of ancient Egypt could be measured in thousand-year periods, mostly because of the relative isolation of the country, the strict government control of the means of transportation, and the monopoly of writing among the quasi-priestly class of scribes. (2) Prior to the Reformation, it was a capital crime in England to translate the Bible into the vernacular, lest the lower classes raise questions about the concept of their "station in life," so fundamental to stability of the medieval society. (3) In the early 19th century in certain regions of the U.S. it was a crime to teach black children or adults the rudiments of

– NOTE T –

THE RATIONALE BEHIND
MULTICOMPLEXIONED WARFARE

Our purpose is to discuss the theoretical advantages afforded by the deliberate use of combinations among the several complexions available to one side of a military engagement.

For a given engagement scenario,[42] the outcome is determined by type and strength of the assets (men, materiel, equipment, organization, training, morale, elan, and many other components) in readiness, available for the battle. The enemy will thus do whatever he can to destroy or degrade these assets *before* they can be effectively engaged. In ancient times, this was called "strategy" or "generalship" as contrasted with the actual conduct of the battle, which was the job of the commander, mostly rated in terms of his ability as a "tactician." Whenever the depth of the battle zone exceeds visual range, the success of strategic operations depends on the ability to locate and identify the targets and to reach them with lethal weapons in

41 (Footnote continued)

  literacy; this limitation was *not* restricted to slaves only. (4) In certain French colonies especially printed catechisms aimed at the native youngsters were extolling the God-ordained superiority of the French "sent to rule and to govern" the natives. (5) Closer to our times, each group, junta, or party having a "coup" or revolution in mind considers its first order of business to seize control of the local TV and broadcast stations. (6) The Ayatollah Khomeini, whatever faith he places in the support from Allah, invariably makes certain that tape recorders and casettes with appropriate playback equipment (all of satanic origin, but in fact manufactured in Japan or under U.S. license) are being made available to his faithful followers.

42 Includes the initial conditions, the intelligence available to all sides, the tactics and priorities of the opposing commanders, and, as all military know only too well, the intervention of chance.

time before the engagement.[43] Under modern battle conditions, *the effectiveness of missiles (or aircraft for that
matter) for the purpose of strategic attrition of the enemy
assets is no longer constrained by the limitations on fire-
power, lethality of warheads, or delivery ranges; it is
dominated by prompt, reliable target acquisition and weapon
assignment.*

It follows that the vulnerable point of the enemy's strategic
efforts is his <u>targeting</u>, or to use a more general term, the
information structure used by the enemy to carry out his
attack. As explained elsewhere,[44] one can enlist such brutal,
but time-honored, techniques as destroying the enemy's
attack weapons, or better, his sensors, communications, and
command centers required for effective targeting, but this
cannot be done overtly in peacetime or in the midst of an
escalating crisis. A far more subtle and efficient technique
for degrading the enemy's targeting is to render his decision
processes involved so complex and time-consuming, as to be
for all practical purposes useless, or even worse.[45]

Specifically, if we deploy our weapons, troops, command and
control systems, logistics, deliberately in such a manner
that the task of efficiently targeting them becomes impos-

---

[43]   Prior to World War I the reach of strategy, in the sense used here, was
essentially artillery range, i.e., measured in tens of kilometers (km) at
best. The introduction of aircraft extended this range to a few hundred,
and recently to a few thousand km; modern missile weapons can cover the
range of a few to a few thousand km, while also reducing the time element to a few seconds or a few minutes—if the missile can be reliably
guided to its intended aimpoint. In addition to accuracy, this involves the
sticky problem of target detection, localization, and identification, and all
this in time to conduct an effective attack. Most major modern "battles"
will not have sharply delimited battle zones; all resources within a given
theater could be directly engaged. The delineation between "strategic" and
tactical employment will be blurred beyond recognition.

[44]   Appendix—Note U; and more extensively, Reference 49.

[45]   This is not a mere stylistic cliche: If, owing to the cumbersome approach
to targeting, the enemy is deterred, that is bad for his side, but he can
always try another day. If on the other hand, still owing to erroneous
targeting, he has irretrievably expended valuable resources to attack the
wrong or nonexisting targets, his situation is obviously worse.

sible, our assets will in fact be protected against attempts at destruction. [46] The central point of our argument is that in an era of intrinsically competent acquisition sensors, almost instant weapon delivery, and of powerfully lethal warheads the degradation of the enemy decision processes is one of the remaining effective approaches to protecting our forces.

This point is so important that it is worth repeating. One of the major elements of survivability of our forces is the degradation of the enemy's decision process *before* (and, of course, at any time *during*) the overt period of hostilities. In more colloquial terms, pouring confusion into the enemy's targeting process is more powerful in protecting our forces than casting steel or pouring concrete; it also happens to be potentially less expensive.

Let us now examine the practical consequences. We are attempting to degrade a decision (i.e., targeting); the means available are rather loosely described as saturation, deception and/or confusion; there is no sharp delimitation or theoretical difference among the processes so described. Saturation means that the input stimuli are too frequent for the enemy to handle. The result of saturation is an increase

---

[46] In everyday parlance, we would enhance their *survivability*. This word should be more explicitly defined since its meaning depends on the scenario context. In the prehostilities stage, the main concern is preemption by surprise attack; the related aspect of survivability is in fact the *immunity* to surprise, i.e., the relative insensitivity to warning. In the initial intense phase, when the adversaries attempt by all possible means to destroy each other's assets *before* the actual engagement, surprise is hardly possible to achieve, but massive "exchanges" are certainly expected to take place. Survivability in this context mostly means favorable exchange ratio, i.e., the resources expended by the enemy must be more valuable (to him) than are those destroyed (to us). When the hostilities reach the *protracted* or *sustained* phase, the combat momentum is much decreased on both sides; the adversaries no longer engage in large-scale maneuvering or redeployment of their resources, but attempt to reduce each other by means of attrition, i.e., exhaustion of resources. "Survivability" in this phase includes, of course, the preceding definitions, but must also be extended to cover characteristics of *endurance* and *retainability*. Both of these suggest the operational availability over long-protracted combat periods, irrespectively (almost irrespectively, anyway) of the enemy efforts and tactics.

in error rate and/or an increase in delays to the point that the decision, even when correct, becomes inoperative or useless. Deception, or more generally manipulation, is to cause the enemy to accept as true and real a signal or a message which in fact is false, and *vice-versa*, in variable combinations. Based on erroneously interpreted inputs, his decisions are (in general) less beneficial in terms of his objectives.

Our weapon designers obviously have not waited for this theory to be enounced in order to start implementing (see Figure T-I). They have conceived just about any possible deployment mode, land-fixed and mobile, road-mobile, hardened underground, sea-surface, submarines, underwater capsules, internal and external carry on aircraft. In addition to the multiple basing modes, locations are widely different; some directly in the vicinity of the battle zone, others prepositioned at a few ten km; still some others at rear echelons or even (conceptually at this point) all the way back in the continental U.S. Mobility always was the strength of military forces; ships are mobile *par excellence* (although perhaps not fast enough for modern missilery); artillery, field air defense and long-range surface-to-surface missiles are becoming increasingly so, in their own vital interest. Major weapon systems, such as the intercontinental strategic missiles, have exhausted (conceptually at least) the whole spectrum of possibilities: first, fixed silo ICBM's, then submarine-carried fleet ballistic missiles, together with air-delivered bombs, air-launched SRAM's, and very soon cruise missiles are or will be part of the inventory. We now seriously argue the need for still further diversification in deployment modes. The MX combines ground mobility with deception, but of course this is only one step; very soon active point defense, and later multi-layered defense, will be proposed to add to the problems of the enemy planner.[47]

[47] Note that this emphasis on the multicomplexioned nature of weapon deployment is aimed at the problems of *strategic* decisions of the enemy. A point-by-point analogy could be drawn from the *tactical* decisions; there again the multiplicity of choices available to our forces, rendered possible by the flexible combinations of penetration/engagement characteristics of their weapons, is intended to confuse the task of the enemy commander.

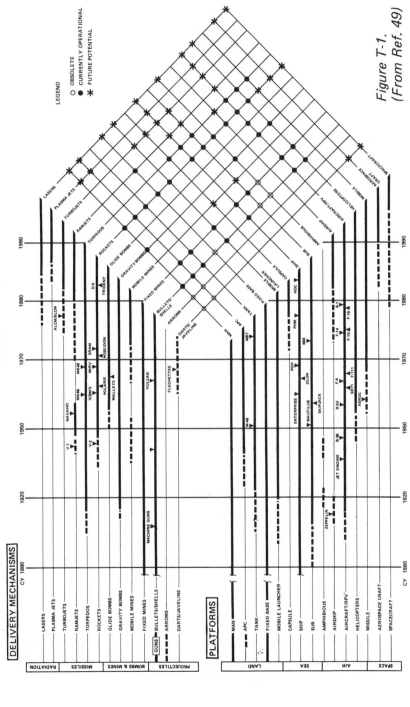

Past and Future

Figure T-1.
(From Ref. 49)

A vastly simplified, but essentially correct way of visuali-
zing the advantages afforded by the multiple-complexioned
deployment concept is to think in terms of combinatorial
analysis. Suppose that the enemy targeting staff is geared
to handle 10 "targets" simultaneously. If we have 10
objects, all identical and time-invariable, as the total inventory of the friendly assets, the enemy's staff problem is
quite tractable. If we now shrewdly allow each of our
objects to assume any of 5 arbitrary complexions, but do not
allow them to interact otherwise, the enemy will have to
contend with 50 simultaneous problems. With a little bit of
effort and overtime, they may still be able to handle it,
especially since not all of our 50 complexions are really
rewarding. Now let us imagine that the same 10 objects
have still 5 time-dependent complexions, and that furthermore 0.01% of their combinations are attractive. The
problem size now is of the order of 1,027,227,[48] which may
just be found somewhat taxing, even for the most devoted
group of ex-komsomols.

---

[48] Combinatorial formula applied to 10 objects, five complexions each,
0.01% of the combinations meaningful.

— NOTE U —

THE INFORMATION WAR CONCEPT[49]
(SUMMARY)

In the recent decades, many spectacular advances have taken place in military technology. Some of these have improved the performance of individual subsystems such as propulsion, structures, guidance, or warheads; some others have ensured the efficient cooperation between all these increasingly complex subsystems by means of integration within the weapon system envelope.

The result of these changes has now been apparent for some time: whenever a weapon can be aimed at its assigned target, the destruction of the latter is assured with a high degree of probability. In the past, protection of targets was often based on passive means such as hardening or on timely counterattacks. There exists now a growing possibility of protecting targets by means of *information denial*. In this protection mode, the defense side aims at depriving the attacker of the essential information required to structure an effective offense. Camouflage, dispersion, and mobility have been used to this effect for many centuries, but modern technology has added strong new impetus to information denial. All forms of warfare, ranging from the highest level nuclear exchange through large-scale "conventional" war to counterinsurgency and guerrilla activity could be impacted.

In many instances, high-performance weapon systems have come to depend critically on interaction with *external* elements, friendly, neutral, or hostile. The command structure, the surveillance and support ancillaries, navigation references, and the target area observables are representative examples. Central to the concept of this *extended weapon* system is the remoteness of the physical elements. Communication links are thus introduced that require a new level of integration; more importantly, they introduce new opportunities for the enemy to practice modern and quite effective versions of information denial. Disruption and manipulation of the adversary's information flow by means of countermeasures have rapidly become some of the most potent means to secure military advantage.

There are numerous confluent technologies at hand to reinforce the belief that information-related counter-measures will further grow in efficacy and sophistication; many new areas of application can be readily envisioned. The basic technology aspects have to do with the theoretical and practical advances in the use of the full electromagnetic spectrum. Transducers are available to transform just about any physical phenomenon into electrical signals, with the attendant capability for transmission, processing, and display for use by human operators. Equipment and soft-ware for rather sophisticated information processing, at rates compatible with the speed and frequency domains of interest to weapon systems, are now available within cost, power, weight, and reliability constraints required to satisfy the demands of the most advanced forms of counter-measures.

In the simplest form, information-flow-related counter-measures attempt to disrupt the communication and infor-mation links of the enemy in the last few moments immedi-ately preceding the detonation of a weapon. Jamming of the command link of a surface-to-air missile is a typical example. It is, however, readily apparent that counter-measures of this type can be applied at many points of a weapon system, covering in fact the whole period of its evolution from development through deployment, mission, and post-mission phases. The spectrum of events pertinent to the information flow, which is the potential target of countermeasures, covers an extremely broad frequency domain: slowly varying strategic intelligence is updated in bursts occurring in a matter of months or years; tactical intelligence, surveillance, or reconnaissance may deal with event durations measured in days or hours; and events related to the terminal engagement can take place in seconds or even microseconds. Countermeasures may address any or all of these frequency domains; they may be concentrated in any one locale, or again dispersed over many elements of the weapon system. In point of fact, they can often be quite successfully applied over protracted time periods without the adversary's specific awareness. The generic set of countermeasures can be defined as comprising the *disruption* of the enemy's information flow, the more intelligent *manipulation* of the hostile information flow and, conversely, all activities aimed at protecting our own systems against those of the enemy.

Technology kindred to that being used to disrupt the enemy's information flow can be applied to the protection of our own. The protection of one's information against counter-measures would be properly termed counter-counter-measures, but there is no real conceptual difference between the two types of operation.

The analysis of the role of countermeasures defined at this level of generality leads to the reexamination of the criteria used to derive weapon system requirements. If, given a set of initial conditions, resources, and available intelligence, two adversaries rationally structure their strategy choices and the corresponding tactical moves, the outcome of the engagement (battle or campaign involving several encounters) is to a large extent governed by the degree of match between the two opposing strategies. How accurately a commander can define his strategy so as to best use the resources available to him depends on the timeliness and accuracy of the information available to him in regard to the enemy resources, intent, and order of battle. *The moves and countermoves related to the information flow, hereafter called information war, are intertwined with, and super-imposed on, other military operations.*[50] They add, therefore, quite a large number of new significant options in the definition of strategies and tactics. Analytical derivation of weapon system requirements in order to "optimize" the outcome of some engagements becomes even less practical than without the consideration of the information-war aspects.

[50] Emphasis added

A modified set of criteria for defining new weapon system requirements can be derived from the insight afforded by the information-war concept:

Weapon systems addressing high-priority missions should be *multicomplexioned*; i.e., having several different and independent means for accomplishing the task. Strategy options attractive to the enemy should be eliminated by avoiding reliance on critical, high-value, and vulnerable elements within our weapon systems that may offer attractive aim points to his counterattacks or entry points for his countermeasures. We should, in the concept development phase of new systems, consciously account for the dynamic aspects of the weapon system development process as impacted by the *informed* responses of the prospective enemy. Our exploratory research aimed at growth options and modifications should address the means for denying to the enemy the developmental moves that may effectively negate the value of our projected new capability.

With a multicomplexioned force, the exercise of tactical flexibility on short notice is possible and highly desirable. The commander of the friendly side should be in position to rapidly modify the nature of his engaged resources and the manner in which his forces are deployed ("order of battle"). Here again, the opportunities offered by manipulating the information flow in the sense of disruption and deception may be of considerable value. If the changes in our engagement posture occur at a faster rate than the enemy's intelligence/reaction cycle, his response will be found to be less than adequate and his chances for success are correspondingly decreased.

Starting from a purely technical observation—the all-pervasive nature of information flow in weapons and combat operations—the conceptual aspects of countermeasures have led us to define the elements of the information war. The possible impact on the outcome of engagements has been assessed on mostly analytical grounds, suggesting a shift in emphasis among weapon system requirement criteria. The problems of disrupting and manipulating the enemy's strategic and tactical intelligence (as well as protecting our own) over the entire development, deployment, and operational phases of weapon systems should attract much increased attention of the defense community.

The reader might be justified in raising a few intriguing questions. Is the information war concept recognized within the U.S. Department of Defense as an essential adjunct to mission and system requirement definition? In the affirmative, how are considerations derived from the information war concept reflected in policies, directives, and procurement procedures without destroying the essential merits of our initiatives or countermoves? How does the information war concept relate to arms limitation talks, including the associated inspection or monitoring systems? *How does an "open" society, with its emphasis on freedom of information and public scrutiny, protect its interests in a hostile world suffused with long-term moves and countermoves of the information war? In particular, how do civilian propaganda and psychological warfare interface with the problems we have discussed?* [51]

As a direct result of this study, we can do no more than hint that these broader questions deserve exploration and that the answers may be of some relevance to our future military posture.

---

[51] Emphasis added

– NOTE V –

STRATEGIC ARMS LIMITATION TREATY (SALT)[52]

The purpose of arms control agreements is to achieve respective military postures satisfactory to all participants at resource expenditure levels significantly lower than those that would have occurred in the absence of such agreements. The key to successful arms control is the reality, as well as the perception, of reduced military threat levels resulting from the arms control constraints.

Negotiated or tacit arms control agreements prohibit, limit, or delay the development, deployment of certain types of weapons.[53] Explicit agreements of this type may also include a number of peripheral clauses related to compliance, such as verification, noninterference with monitoring, nonintrusion under the pretense of verification, and disclosure in case of accidents.[54]

This note is aimed at assessing the results of the U.S./USSR SALT negotiations and agreements to date and at deriving a certain number of recommendations in regard to the future.

As a matter of historical perspective, there is no single instance of an arms control treaty having achieved even partial success in moderating the pace of military competition when the technical basis for superior weaponry was at hand and the required capital resources were available.[55] Specifically, treaties appear to constrain unilaterally Western democracies when they are faced with dictatorships.

---

[52]   Added in proof

[53]   When these are strategic nuclear offense or defense forces, it is customary to refer to the corresponding arms control negotiations and agreements as "Strategic Arms Limitation Treaties" (SALT).

[54]   Disarmament is a special case calling for elimination of specific components of military posture (weapons, bases, etc.) from the operational inventories.

In the specific context of the U.S.-USSR SALT negotiations, it can be categorically stated that they can not possibly result, under the present conditions, in outcomes favorable to Western interests. (1) The characteristics of the two societies are basically different; thus their actions concomitant and following arms control agreements are asymmetrical and demonstrably result in major disadvantages to the West[56]; (2) The substance of current negotiations is such that any misjudgment on a matter of substance could result in catastrophic, irreversible consequences; (3) Essential elements of equivalence or supremacy, as the case may be, are intrinsically unverifiable[57] and thus can not be monitored with any degree of confidence; and (4) Most importantly, in

[55] The "Limitation on Armaments Conference" was held from November 12, 1921, through February 6, 1922, in Washington, D.C. Major powers signed this agreement to outlaw poison gas, to curtail naval construction, and to restrict submarine attacks on merchantmen; also to respect the integrity of China. Especially in regard to this latter clause, subsequent history speaks most eloquently against placing high trust in arms control agreements.

The "London Naval Reduction Treaty" was signed April 22, 1930, by the U.S., Great Britain, Italy, France, and Japan. It went into effect January 1, 1931, and expired December 31, 1936. It constrained only the shipbuilding programs of the U.S., the U.K., and France; none of the signatories were anxious to renew by the time the Second World War was just beyond the horizon. Certainly, Italy and Japan were not constrained; as to Germany, the Third Reich violated even the 1919 Versailles Peace treaty in building up its naval and submarine forces.

[56] In the closed Soviet society, it is possible to undertake developments and even deployments with relatively low probability of prompt and accurate detection. It is also possible to retain technically competent weapon development teams and to exploit the results of "peaceful" nuclear explosions permitted by a hypothetical comprehensive nuclear weapon test ban (CTB). Such action would not be practically possible in the U.S. Our society is essentially transparent to intelligence (except for the highest security levels, perhaps) unless some of the recommendations of Chapter XIII are given serious consideration.

[57] The intrinsic capability of a strategic weapon system depends on minute technical details not easily accessible to intelligence in a closed society. In important instances, the quality and timeliness of intelligence estimates are inadequate for reliable assessment of aggregate strategic capability of the enemy, especially when these involve matters of training, readiness, decision mechanisms and intent.

the ultimate analysis, the success of arms control must rest on shared value judgments that are nonexistent in the present and foreseeable state of antagonism between the two superpowers.   No solemn declarations of politicians, no pleadings of Allies and neutrals, no objurgations by moralists, humanists, and pacifists will change this fact.

If these reasons, based on logic, were insufficient, the empirical evidence in support of pessimism as regards the future SALT prospects is quite substantial.   Since 1968 (start of the SALT process under President Johnson) and more perceptibly since 1972, the U.S. alone felt to be unilaterally constrained by the spirit and the letter of the treaty and of its successors (Vladivostok, 1974 and SALT II 1979, not ratified by the U.S.).   The Soviets may have abided by the letter[58], but most certainly not by the spirit. In point of fact, the combination of SALT constraints and "detente" atmosphere has progressively modified the U.S. strategic posture in relation to the Soviet Union from one of unquestioned supremacy to highly questionable and problematic equivalence.[73]

In order to gain a balanced picture of the SALT phenomenon in the context of the U.S. strategic posture development, it is useful to review all the reasons, true or alleged, why until very recently the SALT negotiations have been considered fundamental to U.S. policy. A few juxtaposed comments on the corresponding Soviet attitudes may be also appropriate.

• Both sides want to project a peace-loving image to their own population and abroad. A well-orchestrated chorus in the West seizes every opportunity for reciting the horrors of nuclear wars; it is part of conventional wisdom that the more weapons that are around, the higher is the probability that nuclear war will in fact

---

[58]   There are serious analysts of the U.S./USSR strategic nuclear balance who maintain that a number of significant "massive operational concealments and ruses" by the Soviets have in fact taken place.[74] There are also reasons to believe that treaty violations continue to occur in the area of encryption of weapon test telemetry data.[75]

occur.[59] The U.S. is particularly anxious to atone for having exploded deliberately two atomic bombs over city populations in 1945. It will go to great length to assure the world that avoidance of nuclear wars is its first national priority... even when this is patently not its acknowledged action policy.

• Before the Soviets reached nuclear parity (and some insist on using the word "superiority"), the U.S. had some illusions about educating the Soviet policy-makers about the more subtle aspects of military strategy applicable to the nuclear age. As it turned out, the Soviet leaders were not particularly in need of military strategy lessons.[76] As evidenced by the results of their single-minded pursuit of nuclear superiority, they might be well justified to offer us at least a valuable case study in negotiating tactics.

• The relationship of arms control, and more particularly of SALT, to other elements of international good behavior is, to say the least, ambiguous in the Western mind.[60] The early linkage theory offered in support of SALT evaporated somewhat during the 1973-78 period; it was considered as unnecessary or even inappropriate in the early Carter years; it came back into vogue in 1979 (in the negative sense, to pressure the Soviets into civilized behavior over the Cuban combat brigade incident, then later in 1980 in regard to the events of

---

59 The probability of accidental nuclear explosion may increase with the number of devices in existence, but this is based on inference without scientific basis in fact. The likelihood of accidental nuclear explosions has probably more to do with the soundness of safing and arming procedures, which may be even better perfected as the funding for weapon development programs becomes adequate, as contrasted to penurious. Be this as it may, the probability of occurrence associated with general nuclear war has no known relationship to the absolute number of nuclear weapons in place; on the other hand, it may have a significant relationship to the state of nuclear balance and stability, real or perceived.

60 The Soviets have no such difficulty; they have long ago decided that detente and SALT are perfectly compatible with supporting wars of national liberation and other anti-Western moves elsewhere. All this, of course, is in keeping with Marxist-socialist dialectics. . .

Afghanistan). [61]   When explicitly questioned, most Administration officials from 1972 to this day would *not* go on record with the statement that risks or disadvantages in the strategic nuclear area could or should be traded against gains in relation to other (nonstrategic-nuclear) policy concerns.

- It is possible, and indeed likely, that the Soviet Union uses SALT as a means to manipulate the U.S. public opinion against a firm stand on nuclear weapon developments.   By offering the appearance of reason, by engaging in the trappings of serious and protracted negotiations, the Soviets manage to delay and to dilute the U.S. strategic nuclear weapon development programs. During the course of negotiations, they may also obtain direct corroboration on characteristics of our own weapon systems in their possession through intelligence channels.

- It used to be a popular pronouncement that the cost savings from reducing expenditures in strategic nuclear weapons may be used for some other purpose, peaceful if possible, but at least some other, morally more acceptable or militarily more traditional undertakings. If the same argument is effectively debated in the Politburo, its echoes have not been reliably detected by Western monitors.   At any rate, the U.S. public has certainly not seen any serious, long-term impact of arms control on the military budget; if anything, the absolutely essential insurance-type developmental programs having been delayed, the U.S. now faces a sizeable "back bill" to pay for its lack of wisdom of the recent past.

- There is a residual possibility that if arms control agreements call for mutual disclosure of weapon characteristics and deployment data, and if the Soviets choose to be truthful in such disclosures, corroboration

[61]   (Note added in proof) The current (April 1981) view of the Reagan Administration seems to seek the results of future SALT negotiations as desirable (or undesirable) independently of any linkage to other foreign policy activities.

by our independent intelligence channels may in the long run calibrate Soviet "good will" and veracity. If further reinforced by the genuine observation of other treaty clauses, this may offer over the long run evidence of (slow) evolution of common values. If it truly materializes, this feature may be the only lasting merit of the SALT process.

On balance it would appear that the only valid reasons wherefore the U.S. may still wish to continue the strategic arms control negotiations are related to peaceful image projection and perhaps to the far more tenuous hope that the Soviet behavior, in terms of veracity and restraint, may justify after a suitably long time period increased U.S. reliance on their treaty commitments. All other reasons, i.e., mutual education, cost savings, reduction of the risk of accidents, are false or no longer operative. In view of the clearly minimal and problematic payoff for the U.S. one is probably justified in suggesting a certain number of *sine qua non* conditions for the conduct of future negotiations.[62]

First, our policy objectives to be achieved should be clearly defined and publicized. Specific proposals must be analyzed from the standpoint whether or not they contribute in a positive and demonstrable way to furthering our objectives. Any provision that is weak or questionable in that regard should be resisted or discarded. Provisions which are strong and supportive of our stated objectives must be emphasized and presented as not negotiable. If for any reason our policy objectives in regard to strategic nuclear forces cannot be constructively served by further negotiations, these should be stopped, abandoned, or at least recessed until new political and economic circumstances warrant reconsideration.

[62] Extracted from Ref. 77 and verbal communication by Mr. B. T. Plymale.

Second, any treaty which the U.S. eventually ratifies should substantially encourage, or compel, the adversaries to invest in systems which are essentially survivable in their pre-launch mode. Specifically:

• The U.S. should insist on inclusion of the mobile land basing mode of ICBM's among those authorized, including redundant deployment modes.

• Anti-ballistic missile defense, applied to ICBM siting, should be permitted, as potentially contributing to launcher survival and therefore to stability.

• Limits should be placed on the number of individual reentry vehicles as contrasted to launchers or plat-forms. On the total ballistic missile throw-weight, limits should be imposed for both ICBM's and SLBM's.

• The U.S. should be allowed to implement substantial diversification, dispersion, and redeployment of its strategic offense forces.

Third, future treaties should impose the obligation on the two parties to disclose information in their possession as to the survivability of their own and of their opponents' strategic weapons. The opponents should be compelled by treaty to disclose all essential information in their possession such as their warhead types and production rates, the location of their fixed missile sites, the number and acoustic signatures of submarines on station, and general deployment areas. U.S. means of independent verification could then be calibrated and could also promptly detect whether the Soviets act in good faith or whether they substantially attempt to hide the truth from our inspection. Consistent, long-term experience with voluntary statements and actions corroborating our information, and positively supporting agreed-upon objectives would alleviate many misgivings associated with the current arms control treaty patterns.

Fourth, the U.S. should decide on a development and deployment program for its strategic offense, defense, and command and control forces. It should comprise the relevant elements of, or interfaces with, other force components such as ASW, theater nuclear weapons, space operations, civil defense and industrial recovery reserves: it should be defined with the assumption that the U.S. strategic forces must ensure reliable deterrence of the Soviets under any practically conceivable crisis scenario. Should deterrence fail, fighting, prevailing in combat, and protecting U.S. and Allied populations and wealth become the prime objectives. *Under no conditions should the U.S. slow down its research, development, production, and acquisition momentum, or reduce the planned outlays on account of arms control treaties being successfully negotiated.* Effective forces-in-being are the only effective negotiating stance the Soviets seem to respect.

References

1. Banfield, Edward C., *The Unheavenly City*, 1968, Little, Brown and Company.
2. Darwin, Charles, *The Origin of Species*, 1859
3. Wilson, E. O., *Sociobiology*, 1975, Belknap Press
4. Heinrich, Bernd, *Bumblebee Economics*, 1979, Harvard University Press
5. McNeil, William H., *The Rise of the West*, University of Chicago Press, 1963
6. Drucker, Peter, *The Age of Discontinuity*, 1978, Harper-Row
7. Toffler, Alvin, *Future Shock*, 1970, Random House
8. Norman, Colin, "Snail Darter's Status Threatened," *Science*, Vol. 212, No. 4496, 15 May 1981, p. 761
9. Carson, Rachael, *Silent Spring*, 1962, Houghton, Mifflin
10. Cohen, B. L., "Disposal of Radioactive Wastes from Fission Reactors," *Scientific American*, July 1977, pp. 21-31
11. Meadows, D. H., et al, *The Limits to Growth: A Report of the Club of Rome Project on the Predicament of Mankind*, 1974, Universe
12. Simon, Julian L., "Resources, Population, Environment: An Oversupply of False Bad News," *Science*, Vol. 208, 27 June 1980, pp. 1431-1437
13. Ormerod, W. E., "Ecological Effect of Control of African Trypanosomiasis," *Science*, 27 February 1976, Vol. 191, No. 4229, pp. 815-821
14. Abercrombie, Thomas J., "Egypt—Change comes to a Changeless Land," *National Geographic*, March 1977, Vol. 151, No. 3, pp. 312-340
15. Symposium, "Energy and the Law," *Mercer Law Review*, Vol. 30, No. 2, Winter 1979
16. Lewis, Harold W., "The Safety of Fission Reactors," *Scientific American*, Vol. 242, No. 3, March 1980, pp. 53-65
17. Sagan, C., et al, "Anthropogenic Albedo Changes and the Earth's Climate," *Science*, Vol. 206, Dec. 21, 1979, pp. 1363-1367
18. Zaleski, C. Pierre, "Breeder Reactors in France," *Science*, Vol. 208, 11 April 1980, pp. 137-144
19. Spival, Jonathan, "France Pursues Drive to Replace Oil Imports with Nuclear Energy," *Wall Street Journal*, February 29, 1980
20. Bergsten, C. Fred, "Coming Investment Wars?", *Foreign Affairs*, October 1974, pp. 135-152
21. Statistics from *Mineral Facts and Problems*, Bicentennial Edition, U.S. Bureau of Mines, U.S. Dept. of the Interior, 1975
22. Jantscher, Gerald R., *Bread Upon the Waters—Federal Aid to the Maritime Industries*, 1975, Brookings Institution
23. Churchill, W. S., *The Second World War*, Houghton Mifflin Co., Boston, Mass., 1948-1953
24. Holden, Constance, "Physicians Take on Nuclear War," *Science*, Vol. 207, March 28, 1980, pp. 1449-1450

25. Bracken, Paul, "Mobilization in the Nuclear Age," *International Security*, Vol. 3. No. 3, Winter 1978/79, pp. 74-93
26. Brown, W. M. and Yokelson, Doris, "Post Attack Recovery Strategies," *Hudson Institute, HI-3100-RR*, Nov. 1980
27. Luttwak, Edward N., "A Critical View of the U.S. Military Establishment," *Forbes*, May 26, 1980, pp. 37-39
28. Crozier, Brian, "Moscow's Strategic Speed-up for 1979," *Soviet Analyst*, Vol. 8, No. 1, 11 January 1979, pp. 1-5
29. Parker, P. J., "Soviet Military Objectives and Capabilities in the 1980's," 20 July 1978
30. Wade, Nicholas, "For the 1980's, Beware All Expert Predictions," *Science*, Vol. 207, 18 January 1980, pp. 287-288
31. Aiken, H., et al, *Prospects for Large Scale Digital Computers*, MIT Digital Computer Course, 1952 (Attribution unverified)
32. Kahn, H., and Bruce-Briggs, B., *Things to Come (Thinking About the 70's and the 80's)*, 1972, The McMillan Co., New York, N.Y.
33. Branscomb, Lewis M., "Information: The Ultimate Frontier," *Science*, Vol. 203, 12 January 1979, pp. 143-147
34. Majno, Guido, *The Healing Hand*, Harvard University Press, 1975
35. Bazelon, David L., "Science, Technology, and the Court," *Science*, Vol. 208, 16 May 1980
36. Elgin, Duane, and Mitchell, Arnold, "Voluntary Simplicity," SRI Report, No. 1004, June 1976
37. Steinberg, E. G., and Yager, J. A., "New Means of Financing International Needs," The Brookings Institution, Washington, D.C., 1978
38. Salisbury, David F., "Defusing Conflicts Caused by Drive for Energy Alternatives," *The Christian Science Monitor*, March 5, 1980
39. Saltykov-Schchedrin, M., "How a Muzhik Fed Two Officials," *Treasury of Russian Life and Humor*
40. Nakamura, Koji, "Who Benefits from Arming Japan?", *The Christian Science Monitor*, April 10, 1981, p. 22
41. "U. N. Yearbook of International Trade Statistics" (1979) (Extrapolated)
42. Fouquet, David L., "Europe Renews Its Links With 57 Former Colonies," *Christian Science Monitor*, November 8, 1979
43. Editorial, "Where Europe and Third World Meet," *Christian Science Monitor*, February 5, 1980
44. Hardin, Garret, "The Case Against Helping The Poor," *Psychology Today*, September 1974, pp. 38-43, 123-126
45. Anon., "Mozambique Turns to the West," *Time*, April 28, 1980
46. Anon., "Libya's Quiet Investments in NATO Countries," *Business Week*, March 26, 1979, pp. 91-92
47. Kolata, Gina Bari, "Attempts to Safeguard Technology Draw Fire," *Science*, Vol. 212, 1 May 1981, pp. 523-526

48. Thomas, Lewis, *The Lives of a Cell*, The Viking Press, 1974
49. Rona, Thomas P., "Weapon Systems and Information War," July 1976
50. Rona, Thomas P., "Weapons at the Dawn of the 21st Century" (To be published October 1982)
51. Gustavson, M. R., "A Dynamic View of Deterrence," *Comparative Strategy*, Vol. 1, No. 3, pp. 169-182
52. Gustavson, M. R., "Evolving Strategic Arms and the Technologist," *Science*, Vol. 109, December 5, 1975, pp. 955-958
53. Bryson, R. A. and Goodman, B. M., "Volcanic Activity and Climatic Changes." *Science*, Vol. 207, March 7, 1980, pp. 1041-1044
54. Smith, R. Jeffrey, "Nonproliferation Policy Challenged," *Science*, Vol. 208, 2 May 1980, p. 478
55. Kerr, R. A., "Prediction of Huge Peruvian Quakes Quashed," *Science*, Vol. 211, February 20, 1981, pp. 808-809
56. Ashby, E., *Perspectives in Biology and Medicine*, Vol. 23, 7 (1979)
57. Naylor, Thomas H., "The U.S. Needs Strategic Planning," *Business Week*, December 17, 1979, pp. 18-19
58. Bazelon, David L., "Risk and Responsibility," *Science*, Vol. 205, 20 July 1979, pp. 277-280
59. Mehrabian, A., *Non-Verbal Communications*, Aldine Publishing Co., Chicago, 1972
60. Miller, G. A., *Non-Verbal Communications in Communications, Language and Meaning, Psychological Perspectives*, edited by Miller, G. A., Basic Books, Inc., New York, 1973
61. *The Information Bank*, (New York Times Co.) May 16, 1980
62. Hardin, G., *Nature and Man's Fate*, 1959, Rinehart & Co., Inc., New York, p. 54
63. Gitto, Paul, "Toward the Automated Factory (Direct Numerical Control Interfacing with Computer Aided Design/Computer Aided Manufacturing)," *Astronautics and Aeronautics*, March 1980, pp. 52-57
64. Editorial, "A 'Balanced' Budget," *The Wall Street Journal*, May 12, 1980
65. Malabre, Alfred, Jr., "Debt Grows at Same Pace as Postwar Economy, But Spurt in Private Borrowing Worries Some," *The Wall Street Journal*, February 26, 1980
66. Editorial, "Cameras and Captives," *Wall Street Journal*, December 12, 1979
67. Forbath, P., *The River Congo*, Harper and Row, N. Y., 1977, pp. 44-46
68. Zweig, S., *Conqueror of the Seas—The Story of Magellan*, McMillan Co., 1938
69. *Europa Year Book 1979—A World Survey*, Vol. 1, Europa Publications Ltd., 18 Bedford Square, London, England
70. Menard, H. W. and Sharman, George, "Scientific Uses of Random Drilling Models," *Science*, Vol. 190, 24 Oct. 1975, pp. 337-343
71. Hardin, Garrett, "The Tragedy of the Commons," *Science*, Vol. 162, No. 3859, 13 December 1968, pp. 1243-1248

72. Huntford, Roland, *The New Totalitarians*, 1972, Stein and Day
73. Plymale, B. T., "The Evolution of U.S. Declaratory Strategic Policy," *Journal of International Relations*, Fall 1977, Vol. II, No. 3
74. Sullivan, David S., "The Legacy of SALT I: Soviet Deception and U.S. Retreat," *Strategic Review*, Winter 1979, pp. 26-41
75. Cooley, John K., "Recent Soviet Missile Test Raises SALT II Questions," *Christian Science Monitor*, February 19, 1980
76. Pipes, Richard, "Why the Soviet Union Thinks It Could Fight and Win a Nuclear War," *Commentary*, Vol. 64, No. 1, July 1977
77. Plymale, B. T., and Rona, T. P., "SALT: A Gloomy Future," *Air Force Magazine*, October 1978, pp. 64-68

Glossary

ABM	Antiballistic Missile
AEC	Atomic Energy Commission
AID	Agency for International Development
ASW	Antisubmarine Warfare
CENTO	Central Treaty Organization
CIA	Central Intelligence Agency
CRAF	Civilian Reserve Air Fleet
DOE	Department of Energy
ECCM	Electronic Counter-Countermeasures
ECE	Economic Commission for Europe
ECM	Electronic Countermeasures
EEC	European Economic Community
EEO	Equal Employment Opportunity
EFTA	European Free Trade Association
EPA	Environmental Protection Agency
FAA	Federal Aviation Administration
FDA	Food and Drug Administration
FEBA	Forward Edge of Battle Area
HEW	Health, Eduction, and Welfare
IA-ICOSOC	Inter-American Economic and Social Council
IANEC	Inter-American Nuclear Energy Commission
IBRD	International Bank for Reconstruction and Development
ICBM	Intercontinental Ballistic Missile
ICC	Interstate Commerce Commission
IDA	International Development Agency
IMF	International Monetary Fund
LDC	Less Developed Countries
MENS	Mission Element Need Statement
MIT	Massachusetts Institute of Technology
MITI	Ministry of International Trade and Industry (Japan)
NACA	National Advisory Council on Aeronautics
NASA	National Aeronautics and Space Administration
NATO	North Atlantic Treaty Organization
NLRB	National Labor Relations Board
NSC	National Security Council
NTSB	National Transportation Safety Board
OAS	Organization of American States

OECD	Organization for Economic Cooperation and Development
OPEC	Organization of Petroleum Exporting Countries
PRC	Peoples Republic of China
SALT	Strategic Arms Limitation Treaty
SAT	Scholastic Aptitude Tests
SEATO	Southeast Asia Treaty Organization
U.K.	United Kindgom
UNCTAD	Conference on Trade and Development (United Nations)
UNESCO	Educational, Scientific, and Cultural Organization (United Nations)
UNIDO	Industrial Development Organization (United Nations)
USIS	United States Information Service
VLCC	Very Large Crude Carrier (Ships)
WPA	Works Projects Administration